ADVANCED CHEMICAL KINETICS

Edited by **Muhammad Akhyar Farrukh**

Advanced Chemical Kinetics
http://dx.doi.org/10.5772/68089
Edited by Muhammad Akhyar Farrukh

Contributors

Muhammad Shahzad, Mahabubur Chowdhury, Dongwon Jung, Thabang Abraham Ntho, James Aluha, Pumeza Gqogqa Gqogqa, Tatiana Duque Martins, Antonio Carlos Chaves Ribeiro, Geovany De Souza, Diericon Cordeiro, Ramon Miranda, Lucas Aguiar, Flavio Colmati, Leandro Lima Carvalho, Renan Reis, Wemerson Santos, Roberto Lima, Rodrigo Costa-Félix, Raphaela Baesso, Pamella Oliveira, Gabriel Cataldo, Andre Alvarenga, Alexander S. Mukasyan, Christopher Eugene Shuck, L. Rajendran, Carlos Fernandez, Chitra Devi Mohan, Qiuming Peng, A.Neren Ökte, Murtaza Sayed, Luqman Ali Shah, Javed Khan, Noor Samad Shah, Rozina Khattak, Hasan Khan, Martin Nielsen

Published by InTech

Janeza Trdine 9, 51000 Rijeka, Croatia

Publishing Process Manager Ana Vais
Technical Editor SPi Global
Cover InTech Design team

First published February, 2018
Printed in Croatia
Legal deposit, Croatia: National and University Library in Zagreb

Additional hard copies can be obtained from orders@intechopen.com

Advanced Chemical Kinetics, Edited by Muhammad Akhyar Farrukh
p. cm.
Print ISBN 978-953-51-3815-0
Online ISBN 978-953-51-3816-7

For EU product safety concerns:
IN TECH d.o.o., Prolaz Marije Krucifikse Kozulić 3, 51000 Rijeka, Croatia,
info@intechopen.com or visit our website at intechopen.com.

Meet the editor

Dr. Muhammad Akhyar Farrukh is an associate professor and the founding director of Nano-Chemistry Lab. at GC University Lahore, Pakistan. He has been awarded three gold medals for his outstanding academic performance in physical chemistry and three gold medals for his excellent performance in research and service to society. He has been awarded many international and national awards: selected by UNESCO as the "Representative of Pakistan" in the first Assembly of the WAYS in Morocco, awarded "Young Chemist Award" by IUPAC in Italy, selected as "Young Scientist" by TWAS in Egypt and IAP in Germany and by IAP/World Economic Forum in China, selected as "Young Researcher" at the 63rd Lindau Nobel Laureate Meeting, awarded the IUPAC-2015 *Award for Chemists* as an outstanding chemist from developing countries in S. Korea, received highest *Research Productivity Awards* (Category A) in 2016 and *Ranked 11th in Pakistan* in Chemistry.

Contents

Preface

The purpose of the book on *Advanced Chemical Kinetics* is to provide insight into different aspects of chemical reactions both at the bulk and nanoscale level. The book covers topics from the basic to advanced level so that readers can get maximum benefits for teaching, learning, and doing research in the area of chemical kinetics.

This book has been divided into three sections, respectively (i) "Kinetics Modeling and Mechanism," (ii) "Kinetics of Nanomaterials," (iii) "Kinetics Techniques," to make it more convenient for readers.

The first section "Kinetics Modeling and Mechanism" covers six chapters with a wide range of topics. It focuses on activation energy and complexity arising during the chemical reaction. The chapters in this section will provide a measurement of reaction routes with detailed mechanism, presented with the help of modern techniques. Mathematical modeling analysis and simulation of enzyme kinetics are discussed in the form of differential equations. These models provide analytical understanding due to their potential in predicting kinetic processes. One of the chapters demonstrates the autoignition and chemical kinetic mechanisms of homogeneous charge compression ignition combustion for the fuels like methane, dimethyl ether, isooctane, and n-heptane as the single-stage and two-stage ignition fuels. A zero-dimensional single-zone engine model of "CHEMKIN" in Chemkin-Pro is used to study mechanisms for high- and low-temperature reactions. Photophysical processes and photochemical changes, presented by new materials and devices to provide a control of energy transfer processes, are discussed in detail. The mechanism of hydroxyl radical, hydrate electron, and hydrogen atom is given with the examples of ciprofloxacin, norfloxacin, and bezafibrate. The last chapter of this section covers the acceptorless alcohol dehydrogenation as an effective tool for organic syntheses for economic production of a wide range of organic compounds, such as aldehydes, ketones, ester, amides, carboxylic acids, etc.

Nanomaterials nowadays are playing an important role in catalysis, degradation of organic pollutants, wastewater treatment, etc. The understanding of the kinetics of nanomaterials is important. To cover this gap in knowledge, the second section of this book is dedicated to "Kinetics of Nanomaterials."

The third section highlights an overview of experimental techniques like electrothermal explosion, differential thermal analysis, electro-thermography, X-ray diffraction, ultrasound, gas chromatography, and hydrogen nuclear magnetic resonance used to study the mechanism of reactions.

I would like to thank Publishing Process Manager Ms. Vais Ana for her cooperation throughout the process of the publication of this book.

Muhammad Akhyar Farrukh
Founding Director/In-Charge
Nano-Chemistry Laboratory
GC University Lahore, Pakistan

Section 1

Kinetics Modeling and Mechanism

Chapter 1

Complex Reactions and Dynamics

Muhammad Shahzad and Faisal Sultan

Additional information is available at the end of the chapter

http://dx.doi.org/10.5772/intechopen.70502

Abstract

Starting from the general idea of reaction kinetics, their classification, concentrations, and chemical equilibrium, we will focus on their activation energy and complexity arising during the chemical reaction. As in complex and higher-dimensional chemical problems, we need special arrangements, specifically, in the case when a system attains different completion paths or several routes. The stiffness of the system can be removed if we distinctly measure their available reaction routes and get a comparison between them and overall reactions. Secondly, the construction and comparison of the invariant region of the manifold based on the modern decomposition techniques in different available reaction routes allow us to discuss the dynamical properties of the system.

Keywords: chemical equilibrium, detailed mechanism, model reduction, reaction routes, invariant manifold

1. Introduction

The chemical kinetics or reaction kinetics is the branch of physical chemistry that deals with the study of chemical processes, their rates, rearrangement of atoms, the effect of various variables, the formation of intermediates, etc. In fact, the chemical kinetics is the study of different factors affecting the speed of a chemical process and gives information about the mechanism of reaction and transition states. At the macroscopic level, the chemical kinetics deals with the study of amount reacted, formed, and the rates of their formation. While at the microscopic or molecular level, we study the mechanism of a chemical reaction, i.e., atomic collision, activation energy at different stages during the reaction.

The chemical kinetics is classified into three types, mathematical, detailed, and applied kinetics, while their elementary reactions are described as unimolecular, bimolecular, and termolecular reactions.

The **mathematical kinetics** deals with the analysis of different mathematical models used in direct and inverse chemical kinetics. These models represent a set of ordinary/partial differential equations and a set of algebraic equations. Further, direct kinetic problems deal with the analysis of steady state or nonsteady state kinetic models consisting of known kinetic parameters. On the other hand, an inverse kinetic problem reconstructs kinetic dependencies and estimates their parameters based on experimental kinetic data, either steady or nonsteady state.

The construction of the mathematical model is the key part of chemical kinetics, which gives a complete description of reaction mechanism and its rates. It provides a working tool to better understand and design chemical processes, i.e., food decomposition and the complex chemistry of biological systems, etc. These models are also used in designing the fast and slow trajectories of complex chemical reactions and modification of chemical reactors to optimize product yield, more efficiently separate products, and eliminate environmentally harmful by-products.

In **detailed kinetics**, we study the reconstruction of detailed mechanism of reaction either based on kinetic or nonkinetic data. These mechanisms consist of a set of elementary steps having forward and reverse reactions along with the governing equation of mass-action law for the kinetic dependencies. In catalyst reactions, it covers the reactant, products, intermediate, surface properties, reaction steps, reaction routes, adsorption properties, etc.

The goal of **applied kinetics** is to study the kinetic dependence of the rate of chemical reactions on their involved or related conditions, i.e., temperature, pressure, concentration, and so on. This dependence can be related to a single or a series of mathematical models usually called kinetic models. These kinetic models are necessary to represent the hierarchy of models at each stage, i.e., initial, intermediate, and final levels as well as to develop an easy way for mathematical simulations of a chemical process.

The reactions in which a single molecule rearranges itself to make one or more products are called unimolecular reactions or a first-order reaction ($A \rightarrow B$), like radioactive decay in which particles are emitted from a single atom. The reactions in which two molecules take part to form a product are called bimolecular reactions or second order ($2A \rightarrow B$ or $A + B \rightarrow C$), like cycloaddition reaction. The reactions in which three particles collide at the same place and time to form a product are called termolecular reactions or third order ($3A \rightarrow B$ or $A + 2B \rightarrow C$). The third-order reactions are not very common as all the three reactants must have to collide simultaneously to form a product.

The chemical reactions in which the reactants are in the same phases are called homogeneous reactions, i.e., the reaction between two gases, two solids, or two liquids. Let us consider a reversible chemical reaction represented as

$$\sum_i \alpha_{\rho i} A_i \underset{k_i^-}{\overset{k_i^+}{\rightleftharpoons}} \sum_i \beta_{\rho i} B_i, \tag{1}$$

here A_i and B_i are the reactants and products, k_i is the rate constants for forward k_i^+ and backward directions k_i^- (that does not depend on the initial concentration of the reactants and products but it does depend on the temperature), and $\rho = 1,\ldots,m$ are the reaction numbers, while the stoichiometric coefficients $\alpha_{\rho i}, \beta_{\rho i}$ are the nonnegative integers.

The reactions in which the reactants are in different phases and their rates are affected by surface areas are called heterogeneous reactions, i.e., the reaction between gases and liquids, solids and liquids, etc. As in the case of gas solid catalytic reactions, reactants at elementary steps will be gas phase component or surface intermediate. Thus, Eq. (1) can now be written as

$$\sum_i \alpha_{\rho i} A_i + \sum_j \alpha_{\rho j} X_j \underset{k_i^-}{\overset{k_i^+}{\rightleftharpoons}} \sum_i \beta_{\rho i} B_i + \sum_j \beta_{\rho j} Y_j, \tag{2}$$

here again A_i and B_i are the reactant and products in the gas phase and X_j and Y_j are the surface intermediate. In a more typical form, it can be written as $\alpha A + \sum_j \alpha_{\rho j} X_j \underset{k_i^-}{\overset{k_i^+}{\rightleftharpoons}} \beta B + \sum_j \beta_{\rho j} Y_j$.

With an assumption that α and β are either zero or one, it implies that only one molecule in an elementary reaction from the gas phase reacts or zero at all.

The concentration of the involved species can be measured as (single step reaction);

At initial space:	a	0
time t > 0:	$(a-x)$	x

The reaction rates measured on either side are $k_i^+(a-x)^{\alpha_{\rho i}}$ (forward rate of reaction) and $k_i^-(x)^{\beta_{\rho i}}$ (backward rate of reaction) and the product formation is the difference between the rate of forward and backward reactions, i.e.,

$$\frac{dx}{dt} = k_i^+(a-x)^{\alpha_{\rho i}} - k_i^-(x)^{\beta_{\rho i}}. \tag{3}$$

The equilibrium is a dynamical process, and when a system goes to an equilibrium, the left-hand side will become equal to the right-hand side, i.e., $\frac{dx}{dt} = 0$

$$k_i^+(a-x)^{\alpha_{\rho i}} - k_i^-(x)^{\beta_{\rho i}} = 0, \tag{4}$$

while the conversions of products to reactants and reactants to products are still going on, although there is no net change in the number of reactant and product molecules.

According to **Le-Chatelier's principle**, if a system at equilibrium state is disturbed by an external force, then the system tries to offset the force and attains a new position.

The system becomes complex when a reaction undergoes more than one pathways when more than one products are formed from the same reactants or different reactants produce the same products. Such types of reactions are called parallel reactions or side reactions, i.e.,

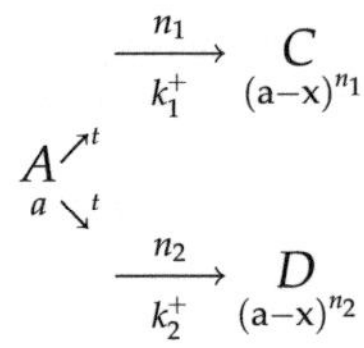

Here a is the initial concentration of the species A and after some time $t > 0$, it dispersed into C and D. Similarly, sometimes it happens that different chemical species give the same products

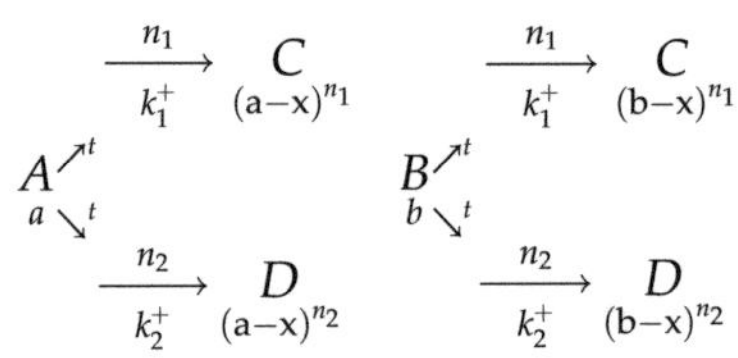

or a system is reversible at different stages. In all these cases, we need to follow all their paths to get the detailed mechanisms,

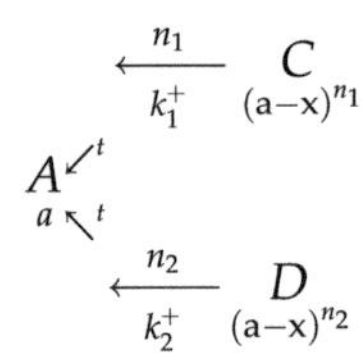

The rate of reaction ($W_\rho(c)$) is proportional to the number of collisions per unit time between the reactants but only a small fraction of the total is effective, i.e., not every collision between the reactants gives the result.

There may be a few reasons behind its ineffectiveness [1], i.e.:

The reactant molecules may attain insufficient energy ($<E_{Act}$, i.e., activation energy, J/ mol) at different stages during the reaction.

The molecules may not get aligned properly or orientate during the collision (depending upon the geometry of the particles and kind of reaction that is taking place), etc.

If Z is the effective collision in which molecules have energy $\geq E_{Act}$, then

$$Z_E = \frac{Z_0 e^{E_{Act}}}{RT}, \tag{5}$$

and $e^{E_{Act}}/RT$ gives the fraction of collisions with energy $\geq E_{Act}$.

In a complex chemical reaction, the reactant molecules (intermediates, complex Θ_i) pass through different transition states due to their bond breaking and energy redistribution factors. Here the species stays for a very short period, usually called transition period of activated complex (where the hidden reactions between the chemical species are still going on very fast).

The energy required to pass the reactant E_R to activated complex E_Θ is called activation energy or energy of activation E_{Act}:$E_\Theta - E_R$. It may be supplied in any form, mechanical, chemical, or thermal, to enable the reactant to convert into the product, i.e.,

$$\sum_{i=1}^{N} \alpha_{S_i} A_i \mapsto \widehat{E_{Act}^F} \mapsto \sum_{i=1}^{N} \beta_{S_i} A_i \tag{6}$$

The activation energy during the forward E_{Act}^F and backward E_{Act}^B reactions must be the same or different depending on the type of reactions. In thermos, the neutral reaction and $\Delta H=0$, the energy of activation in both the directions are same. While in endothermic reactions, $E_{Act}^F > E_{Act}^B$, holds and in exothermic reactions, $E_{Act}^F < E_{Act}^B$. It is also understood that the higher the activation energy, the slower the reaction.

The activated complex is a separate entity and there exists an equilibrium between reactants (products, under reversible reactions) and activated complex (**Figure 1**). Thus, a reaction mechanism can be defined as

$$\sum_{i=1}^{N} \alpha_{S_i} A_i <=> \Theta_i => \sum_{i=1}^{N} \beta_{S_i} A_i \tag{7}$$

But still, there is some activation going on between the activated complexes $\Theta_i^{\mp}$, i.e.,

$$\Theta_i^{\mp} : \Theta_i^R => \Theta_i^P.$$

Therefore, a complete complex chemical reaction mechanism can be defined as $\sum_{i=1}^{N} \alpha_{S_i} A_i \Leftrightarrow \Theta_i^{\mp} \Rightarrow \sum_{i=1}^{N} \beta_{S_i} A_i$.

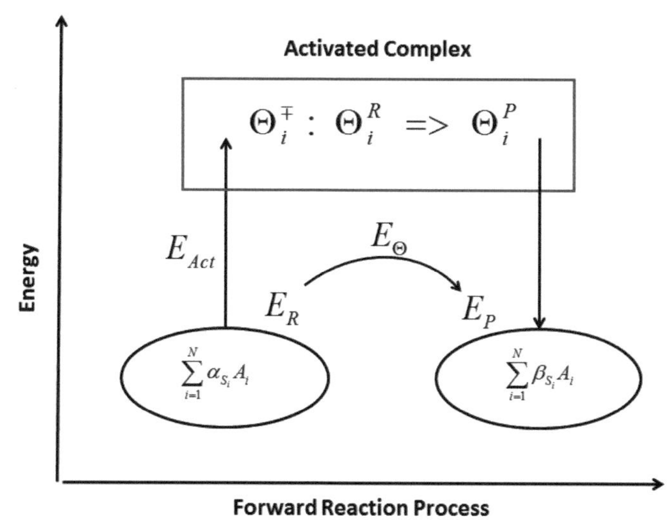

Figure 1. A complex reaction mechanism involving energy barriers and transition states.

In case of reversible complex chemical reactions,

$$\sum_{i=1}^{N} \alpha_{S_i} A_i \Leftrightarrow \Theta_i^{\mp} \Leftrightarrow \sum_{i=1}^{N} \beta_{S_i} A_i \tag{8}$$

The concentration of activated complex can thus be obtained by applying the equilibrium conditions, i.e.,

$$\left[\Theta_i^{\mp}\right] = k_i^{\mp} \sum_{i=1}^{N} \alpha_{S_i} A_i \tag{9}$$

where $\mp$ refers to the activated complex.

2. Reaction rate

A stoichiometric vector γ_ρ of the reaction mechanism (1) is an $n-$ dimensional vector with coordinates $\gamma_{\rho i} = \beta_{\rho i} - \alpha_{\rho i}$, that is, "gain minus loss" in the ρth elementary reaction. In matrix form, it takes a form

$$S = \left[\gamma_{\rho 1}, \ \gamma_{\rho 2}, \dots, \gamma_{\rho i}\right] \tag{10}$$

The chemical composition of the substances is given by the molecular matrix M, with the element m_{ij} as a number of atoms of the jth element in the ith component. M is a $(N_c \times N_e)$ matrix, while N_c is the number of reacting components lying in the mixture consisting of N_e; the chemical elements and the law of conservation of atoms say

$$Mn_c = V_c \ \text{(constant vector)}. \tag{11}$$

The total number of any moles of c_i atoms can be measured by using the relation

$$M^T n_c = n_e \tag{12}$$

Here, M^T is the transposed molecular matrix and n_c and n_e are the component amount (mol) and the amount of the chemical elements (mol) in the column vector form.

The dynamics of the involved concentration species can be measured when we measure the rate of formation of the products or deformation and disappearance of the reactants.

Finally, the rate of reaction will take a form

$$X(c) = \dot{c} = \frac{dc}{dt} = S_\rho W_\rho(c) \tag{13}$$

Here $W_\rho(c)$: $W_\rho(c) = k_\rho(T) \prod_i c_i^{\alpha_{\rho i}}$ is the reaction rate function of the ρth step (i.e., the difference between the rate of forward $W_\rho^+(c)$ and backward $W_\rho^-(c)$ reactions).

http://dx.doi.org/10.5772/intechopen.70502

3. Linear algebra and graph theory

In chemical engineering, the mathematical methods of graph theory have found wide applications in complex chemical reactions and in a sequence of uni (or multi) or parallel reacting events. A graph is a combination of nodes (points) and edges (lines) [2], while a cyclic graph involves finite sequences of edges with the single node (from where it begins and ends).

Similarly, related to any combination of reaction, a tree can be defined as a sequence of noncyclic graph edges. In a spanning tree, certain intermediate may form from other intermediates after a sequence of transformations but does not agree to counter any two reactions with the same step (e.g., +1 and −1) nor two reactions started with the same intermediates (e.g., −1 and +2, or +1 and −3),

Spanning trees can be described in terms of "forward" (generated by a sequence of forwarding reactions), "backward" (generated by a sequence of reverse reactions), and "combined" spanning trees (generated by a sequence of both forward and backward reactions). A single-route, n-steps N_s (edges) reaction mechanism has $N_{\mathbf{int}}$ (intermediates) nodes, such as $N_{\mathbf{int}} = N_{\mathbf{s}} = N$. The total numbers of spanning trees are N^2 in any reaction, while the forward N^f and backward N^b spanning trees are N and the numbers of combined spanning trees N^c are

$$N^c = N(N-2) = N^2 - 2N \tag{14}$$

In a chemical reaction, the overall reaction can be found by multiplying the reactions with certain coefficients, the so-called Horiuti numbers σ, and then adding the results. While the relation between σ and $N_{\mathbf{int}}$ is

$$\sigma.N_{\mathbf{int}} = 0 \tag{15}$$

Horiuti number allows us to distinguish the short-lived intermediate and long-lived components, i.e., to eliminate the intermediates using an RREF of the stoichiometric matrix S, the intermediates must be listed first, not last. Then the rows in which all intermediates vanish provide a basis for the overall reactions [2].

The numbers of key components N_{kc} are given by the equation

$$N_{kc} = N_c - rank(M) \tag{16}$$

and the number of key components equals the number of key reactions. Also, the number of key components + number of nonkey reactions = number of reactions

In **Figure 2**, their curves represent two different solution curves of their respective reaction routes lying at different phase space, i.e., one lies in 2D while the second lies within 3D.

Now the question arises, if a complex reaction adopts different completion routes before giving the product, then how can one relate (or distinguish) such available routes and why they are important to be measured?

For this, the reaction route N_{rr} of the system can be measured as

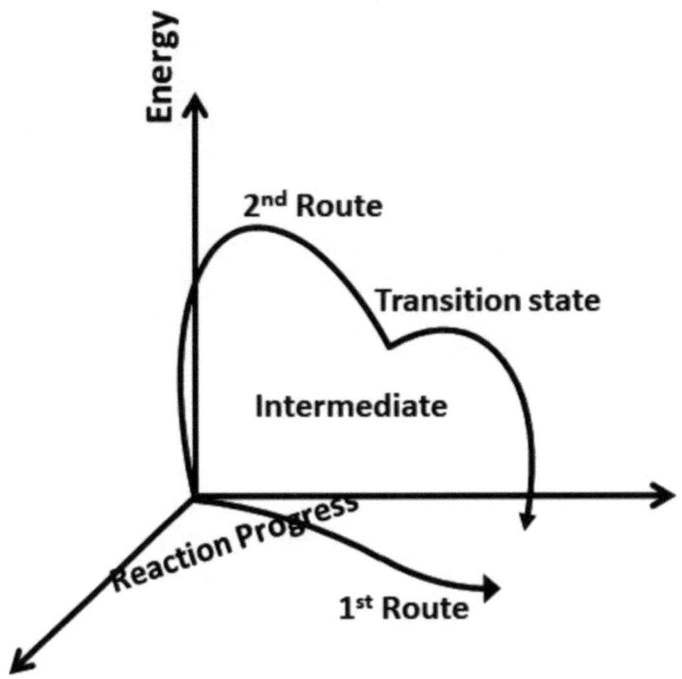

Figure 2. A complex chemical reaction passes through different transition states and adopts different completion routes before giving the products.

$$N_{rr} = N_s - N_{\text{int}} + N_{as} \tag{17}$$

whereas, N_s is the number of steps in the detailed mechanism and N_{as} is the number of active sites in the mechanism. Based on the molecular matrix, the molar masses of the components can be determined from the atomic masses of the elements. The product of the stoichiometric matrix S and molecular matrix M gives

$$SM = 0 \tag{18}$$

To answer the second part of the above question, we need to consider all its available routes to get the detailed reaction mechanism. Then a comparison of these route solutions with the whole reaction mechanism allows us to give any concluding remarks, but we believe that the result obtained through different routes may be similar or vary depending on the type of reactions.

4. Multiroute reactions mechanism

To understand this idea, let us discern the **four-step reversible complex chemical reaction** [3] defined over a closed system having two available routes. The mechanism involves six chemical substances (species C_i) represented as (**Figure 3**),

Steps	Chemical Reaction	σ_1	σ_2
(1)	$C_{A_2} + 2C_Z \underset{k_1^-}{\overset{k_1^+}{\rightleftharpoons}} 2C_{AZ}$	1	1
(2)	$C_B + C_Z \underset{k_2^-}{\overset{k_2^+}{\rightleftharpoons}} C_{BZ}$	0	2
(3)	$C_{AZ} + C_{BZ} \underset{k_3^-}{\overset{k_3^+}{\rightleftharpoons}} C_{AB} + 2C_Z$	0	2
(4)	$C_B + C_{AZ} \underset{k_4^-}{\overset{k_4^+}{\rightleftharpoons}} C_{AB} + C_Z$	2	0

Figure 3. Four-step reversible reaction having two routes.

While the overall reaction evolves no intermediates, i.e., $A_2 + 2B \underset{k^-}{\overset{k^+}{\rightleftharpoons}} 2AB$, the stoichiometric matrix (10) and molecular matrix (11) infer

$$S = \begin{bmatrix} -1 & -2 & 2 & 0 & 0 & 0 \\ 0 & -1 & 0 & -1 & 1 & 0 \\ 0 & 2 & -1 & 0 & -1 & 1 \\ 0 & 1 & -1 & -1 & 0 & 1 \end{bmatrix}, \quad M = \begin{pmatrix} A & Z & B \\ 2 & 0 & 0 \\ 0 & 1 & 0 \\ 1 & 1 & 0 \\ 0 & 0 & 1 \\ 0 & 1 & 1 \\ 1 & 0 & 1 \end{pmatrix} \begin{cases} A_2 \\ Z \\ AZ \\ B \\ BZ \\ AB \end{cases} \tag{19}$$

Note that the relation between them is orthogonal, i.e., Eqs. (15) and (18) hold. While the stoichiometric matrix of intermediates and Horiuti matrix are $N_s = 4$, $N_{\text{int}} = 3(Z, AZ, BZ)$, $N_{as} = 1(Z)$,

$$N_{\text{int}} = \begin{bmatrix} -2 & 2 & 0 \\ -1 & 0 & 1 \\ 2 & -1 & -1 \\ 1 & -1 & 0 \end{bmatrix}, \sigma = \begin{bmatrix} -1 & 0 \\ 0 & 1 \\ -2 & -1 \\ 0 & 1 \end{bmatrix}$$

Atomic balance constraints are given by Eq. (11)

$$\begin{aligned} \eta_{t,A} &= 2\eta_{c_1} + \eta_{c_3} + \eta_{c_6}, \\ \eta_{t,Z} &= 2\eta_{c_2} + \eta_{c_3} + \eta_{c_5}, \\ \eta_{t,B} &= 2\eta_{c_4} + \eta_{c_5} + \eta_{c_6}. \end{aligned} \tag{20}$$

whereas, $n_{t,A}$, $n_{t,Z}$, and $n_{t,B}$ are the total number of moles of A, Z, and B atoms, respectively.

The key components N_{kc} and reaction route N_{rr} of the system are given by Eqs. (16) and (17)

$N_{kc} = N_c - rank(M) = 6 - 3 = 3,$

This means we can reduce this system into three components

$N_{rr}=N_s-N_{\text{int}}+N_{as}=4-3+1=2.$

Hence, this reaction mechanism has two independent routes N_{rr}. Also, when we multiply step-1 and step-3 by its Horiuti numbers, all the intermediates vanished and we get an overall reaction. The same is the case with steps 1, 2, and 4. The dimension of these two routes can be determined by their respective Horiuti numbers. Sets of Horiuti numbers for the first route and second route are (1,0,0,2) and (1,2,2,0), respectively. This implies both the routes are nonlinear.

First-route: two-step mechanisms (**Figure 4**).

Second-route: three-step mechanisms (**Figure 5**).

Steps	Chemical Reaction	σ_1
(1)	$C_{A_2}+2C_Z \underset{k_1^-}{\overset{k_1^+}{\rightleftharpoons}} 2C_{AZ}$	1
(4)	$C_B+C_{AZ} \underset{k_4^-}{\overset{k_4^+}{\rightleftharpoons}} C_{AB}+C_Z$	2

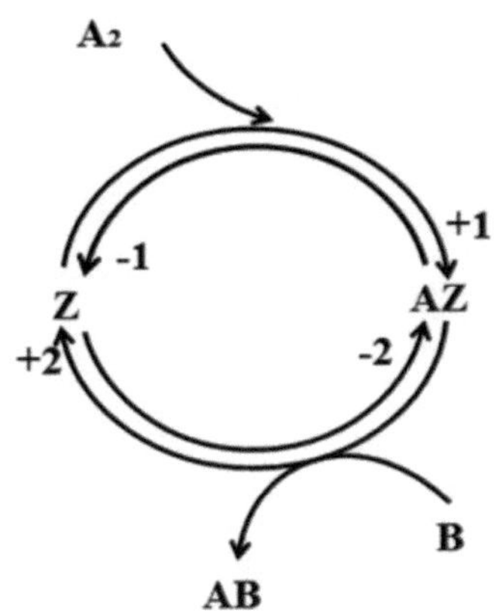

Figure 4. R-1: The first route of the reaction mechanism is a two-step reversible reaction involving five chemical species, while $N_{kc}=2$.

Steps	Chemical Reaction	σ_2
(1)	$C_{A_2}+2C_Z \underset{k_1^-}{\overset{k_1^+}{\rightleftharpoons}} 2C_{AZ}$	1
(2)	$C_B+C_Z \underset{k_2^-}{\overset{k_2^+}{\rightleftharpoons}} C_{BZ}$	2
(3)	$C_{AZ}+C_{BZ} \underset{k_3^-}{\overset{k_3^+}{\rightleftharpoons}} C_{AB}+2C_Z$	2

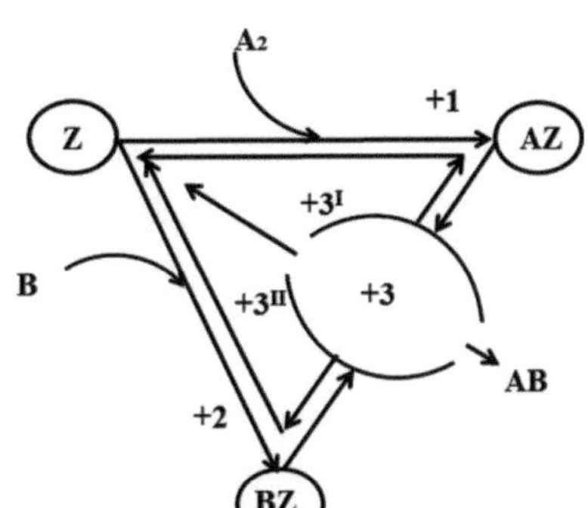

Figure 5. R-2: The second route of the reaction mechanism is a three-step reversible reaction involving six chemical species, while $N_{kc}=3$.

5. The measuring methods

The kinetic equations of the above reaction mechanism (**R-1**) can be measured by using Eq. (13)

$$\frac{d}{dt}\begin{bmatrix} c_{A_2} \\ c_B \\ c_{AB} \\ c_Z \\ c_{AZ} \end{bmatrix} = \begin{bmatrix} -k_1^+ c_1 c_2^2 - k_1^- c_3^2 \\ k_2^+ c_3 c_4 - k_2^- c_2 c_5 - 2k_1^+ c_1 c_2^2 + 2k_1^- c_3^2 \\ -k_2^+ c_3 c_4 + k_2^- c_2 c_5 + 2k_1^+ c_1 c_2^2 - 2k_1^- c_3^2 \\ -k_2^+ c_3 c_4 + k_2^- c_2 c_5 \\ k_2^+ c_3 c_4 - k_2^- c_2 c_5 \end{bmatrix}$$

while N_s=2, N_{int}=2(Z,AZ),N_{as}=1(Z), and a reaction route N_{rr} is 1.

A reduced form of the system (**R-1**) can be achieved by using the Eqs. (12) and (13). While initial parameters are defined as

$$c_1^{eq} = 0.5, c_2^{eq} = 0.1, c_3^{eq} = 0.1, c_4^{eq} = 0.4, c_5^{eq} = 0.1, k_1^+ = 1, k_2^+ = 0.5.$$

Similarly, (**R-2**) implies that we can reduce this system into three components, while N_s = 3, N_{int} = 3(Z,ZO,ZCO),N_{as} = 1(Z), and N_{rr} = 1.

Thus, a single reaction route is available. The kinetic equations for the involved species are given by Eq. (13)

$$\frac{d}{dt}\begin{bmatrix} c_{A_2} \\ c_B \\ c_{AB} \\ c_Z \\ c_{AZ} \\ c_{BZ} \end{bmatrix} = \begin{bmatrix} -k_1^+ c_1 c_2^2 - k_1^- c_3^2 \\ 2k_3^+ c_3 c_4 - 2k_3^- c_2^2 c_6 - k_2^+ c_2 c_5 + k_2^- c_4 - 2k_1^+ c_1 c_2^2 + 2k_1^- c_3^2 \\ -k_3^+ c_3 c_4 + k_3^- c_2^2 c_6 + 2k_1^+ c_1 c_2^2 - 2k_1^- c_3^2 \\ -k_3^+ c_3 c_4 + k_3^- c_2^2 c_6 + k_2^+ c_2 c_5 - k_2^- c_4 \\ -k_2^+ c_2 c_5 + k_2^- c_4 \\ k_3^+ c_3 c_4 - k_3^- c_2^2 c_6 \end{bmatrix}$$

By using Eqs. (12) and (13), a system can be reduced into three numbers of species and their initial parameters are defined as

$$\begin{aligned} &c_2^{eq} = 0.1, c_3^{eq} = 0.1, c_5^{eq} = .4, \ \ c_1^{eq} = 0.500000000 c_3^{eq} c_3^{eq} / c_2^{eq} c_2^{eq}, \\ &c_4^{eq} = 0.25000000000 c_2^{eq} c_5^{eq}, \ \ c_6^{eq} = 0.2500000000 c_3^{eq} c_5^{eq} / c_2^{eq}. \\ &k_1^+ = 1, k_2^+ = 0.5, k_3^+ = 0.5. \end{aligned}$$

Figures 6 and **7** clear the idea of the slow invariant manifold (SIM), i.e., decomposing the system into their fast and slow motion. Their solution trajectories (during their relaxation time) quickly move toward the low-dimensional manifold and after that start moving along it [4–13]. That is the easy way of getting an idea of the SIM. Otherwise, by using the different available methods of SIM, i.e., [14–27] we will get their initial approximations lying on it or near to it. For comparison, we refer the readers to [28].

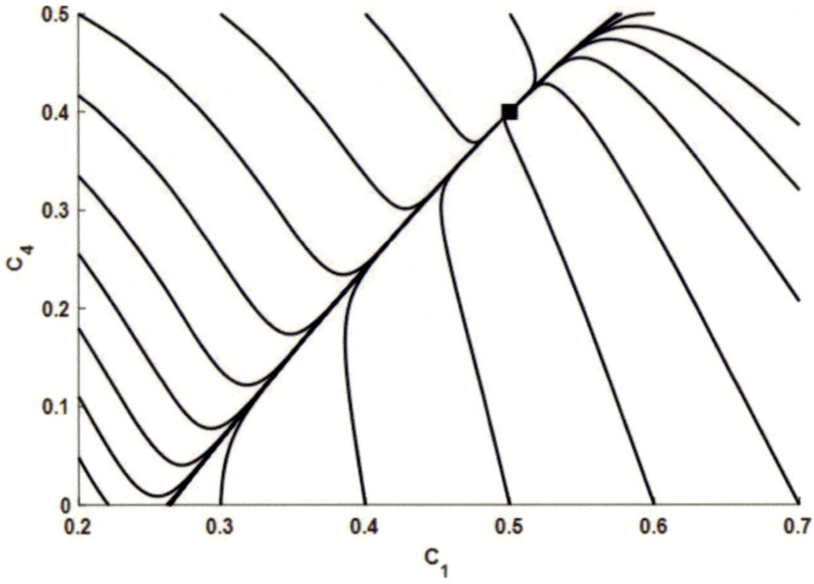

Figure 6. The behavior of the reduced species c_1 and c_4 near the equilibrium point (square). While their solution trajectories approaching toward the equilibrium (during their path) give the region where slow invariant manifold (SIM) lies.

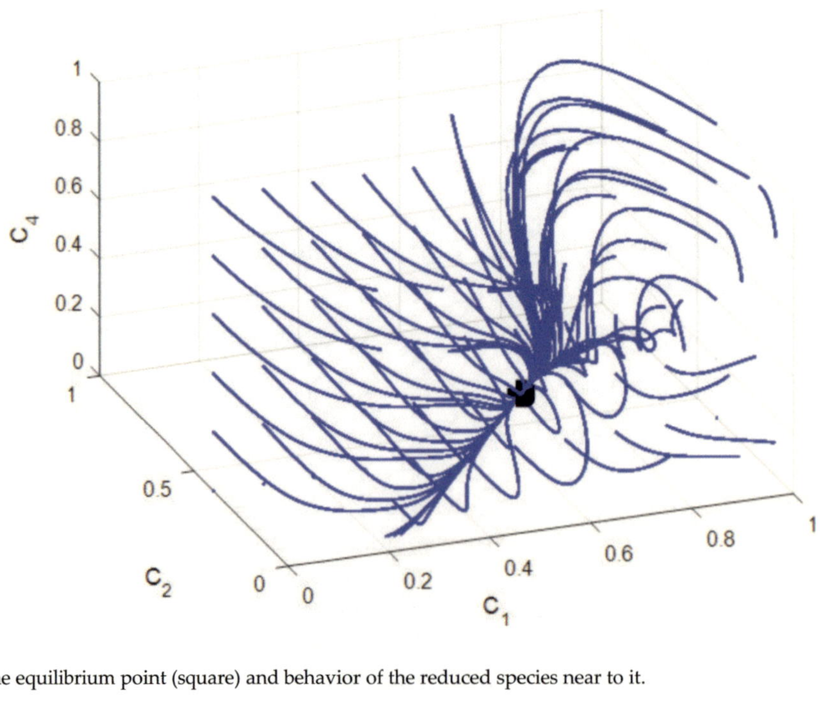

Figure 7. The equilibrium point (square) and behavior of the reduced species near to it.

http://dx.doi.org/10.5772/intechopen.70502

6. The routes comparison

In **Figure 9**, the curves lie in the plane c1 and c4 are not the projected image of the above curves. Instead, it is the behavior of the species measured near the equilibrium point in the first reaction route mechanism, whose $N_{kc} = 2$. Similarly, the above lines describe the behavior of the

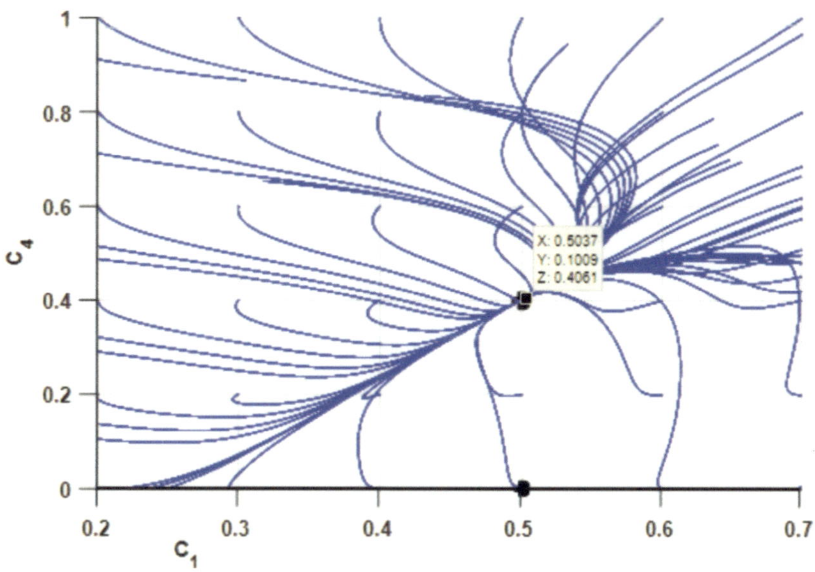

Figure 8. Two-dimensional view of both the reaction routes solutions and both squares represent their equilibrium point.

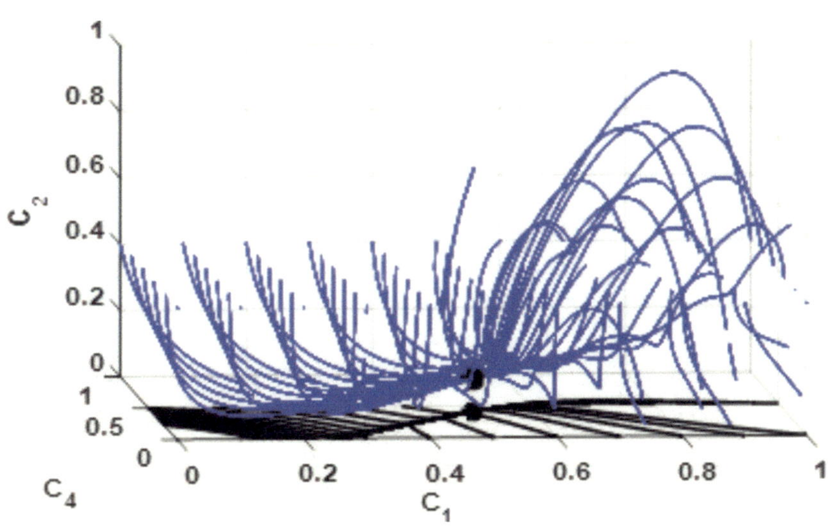

Figure 9. First-route vs. second-route. The three-dimensional view of both the reaction routes solutions and both squares represent their equilibrium point.

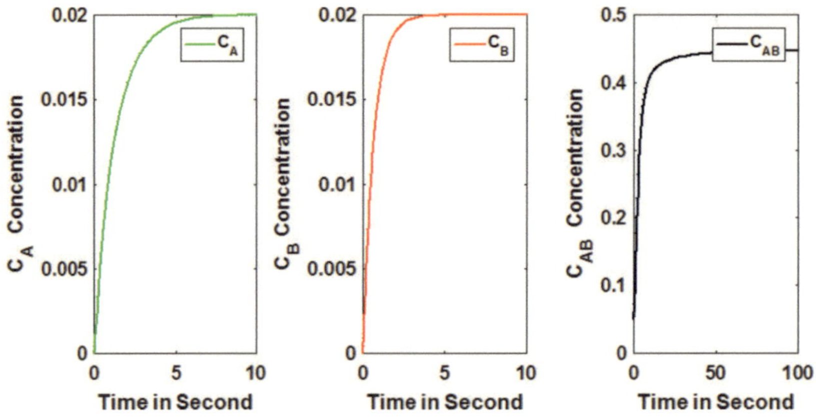

Figure 10. Variation of chemical species concentration of overall reaction mechanism with respect to time, while N_{kc}=1.

species near the equilibrium point measured in second reaction route mechanism, whose N_{kc} = 3. Note that it's invariant region and equilibrium point exactly lie over the invariant region of the first route, i.e., **Figures 8, 9**.

Now, the overall reaction mechanism involves no intermediate, and the variations of the concentration of involved chemical species are given in **Figure 10**.

7. Summary

In this chapter, both the physicochemical conceptual assumptions (used for species behavior and activated complex) and a set of mathematical tools (for their dynamical behavior and simplification) are presented. Mathematically, simplification can be done by "model reduction," that is, the rigorous way of approximating and representing a complex model in simplified form.

Here, we have considered a complex problem having a common step: conferred their available routes then allied graphically. Although we have not applied any numerical or analytical technique to measure the SIM but one can easily examine (by applying such techniques) that their solution trajectories will also lie in the same invariant regions that can also be correlated with each other and even with the whole reaction mechanism.

Thus, the idea initiated here can easily be correlated with the method used for the construction of slow manifold in a complex chemical reaction based on the decomposition techniques of entropy maximum along with certain constraints (lies on the manifold or given by slowest eigenvectors) at the equilibrium point. This will allow us to bring together different available mathematical ideas and methods, commonly used to transform the complex chemical problems from one way to the other, to enhance progress in understanding.

Author details

Muhammad Shahzad* and Faisal Sultan

*Address all correspondence to: shahzadmaths@hu.edu.pk

Department of Mathematics and Statistics, Hazara University, Mansehra, Pakistan

References

[1] Houston PL. Chemical Kinetics and Reaction Dynamics. 3rd ed. Courier Corporation/McGraw-Hill; 2012. p. 352

[2] Constales D, Yablonsky GS, D'hooge DR, Thybaut JW, Marin GB. Advanced Data Analysis and Modelling in Chemical Engineering. 2nd ed. Elsevier; 2016. p. 414. DOI: 10.1007/s11144-017-1163-5

[3] Marin GB, Yablonsky GS. Kinetics of Chemical Reactions. 1st ed. John Wiley & Sons; 2011. p. 428

[4] Al-Khateeb AN, Powers JM, Paolucci S, Sommese AJ, Diller JA, Hauenstein JD, Mengers JD. One-dimensional slow invariant manifolds for spatially homogenous reactive systems. The Journal of Chemical Physics. 2009;**131**(2):024118. DOI: http://dx.doi.org/10.1063/1.3171613

[5] Chiavazzo E, Gorban AN, Karlin IV. Comparison of invariant manifolds for model reduction in chemical kinetics. Communications in Computational Physics. 2007;**2**(5):964-992

[6] Gorban AN, Karlin IV. Invariant Manifolds for Physical and Chemical Kinetics. New York: Springer; 2005. p. 469-489. DOI: 10.1007/b98103

[7] Shahzad M et al. Measuring the complex behavior of the SO_2 oxidation reaction. Computational Ecology and Software. 2015;**5**(3):254-270

[8] Shahzad M et al. Initially approximated quasi equilibrium manifold. Journal of the Chemical Society of Pakistan. 2015;**37**(2):207-216

[9] Al-Khateeb, Ashraf, et al. Calculation of slow invariant manifolds for reactive systems. In: 47th AIAA Aerospace Sciences Meeting including The New Horizons Forum and Aerospace Exposition. January 08; 2009.

[10] Chiavazzo E, Karlin IV. Quasi-equilibrium grid algorithm: Geometric construction for model reduction. Journal of Computational Physics. 2008;**227**(11):5535-5560. DOI: 10.1016/j.jcp.2008.02.006

[11] Shahzad M et al. Slow manifolds in chemical kinetics. Journal of the Chemical Society of Pakistan. 2016;**38**(5):828-835

[12] Kooshkbaghi, M, et al. The global relaxation redistribution method for reduction of combustion kinetics. The Journal of Chemical Physics. 2014;**141**(4):044102. DOI: dx.doi.org/10.1063/1.4890368.

[13] Constales D, Yablonsky GS, Marin GB. Thermodynamic time invariances for dual kinetic experiments: Nonlinear single reactions and more. Chemical Engineering Science. 2012; **73**:20-29

[14] Bongers H, Van Oijen JA, De Goey LPH. Intrinsic low-dimensional manifold method extended with diffusion. Proceedings of the Combustion Institute. 2002;**29**(1):1371-1378. DOI: doi.org/10.1016/S1540-7489(02)80168-7

[15] Bykov V et al. On a modified version of ILDM approach: Asymptotic analysis based on integral manifolds. IMA Journal of Applied Mathematics. 2006;**71**(3):359-382. DOI: doi.org/10.1093/imamat/hxh100

[16] Gorban AN, Karlin IV. Thermodynamic parameterization. Physica A: Statistical Mechanics and its Applications. 1992;**190**(3):393-404. DOI: doi.org/10.1016/0378-4371(92)90044-Q

[17] Gorban AN, Shahzad M. The Michaelis-Menten-Stueckelberg theorem. Entropy. 2011; **13**(5):966-1019. DOI: 10.3390/e13050966

[18] Maas U, Pope SB. Simplifying chemical kinetics: intrinsic low-dimensional manifolds in composition space. Combustion and Flame. 1992;**88**(3):239-264. DOI: doi.org/10.1016/0010-2180(92)90034-M

[19] Gorban AN, Karlin IV, Zinovyev AY. Invariant grids for reaction kinetics. Physica A: Statistical Mechanics and its Applications. 2004;**333**:106-154. DOI: doi.org/10.1016/j.physa.2003.10.043

[20] Gorban AN, Karlin IV. Method of invariant manifold for chemical kinetics. Chemical Engineering Science. 2003;**58**(21):4751-4768. DOI: doi.org/10.1016/j.ces.2002.12.001

[21] Gorban AN, Karlin IV, Zinovyev AY. Constructive methods of invariant manifolds for kinetic problems. Physics Reports. 2004;**396**(4):197-403. DOI: doi.org/10.1016/j.physrep.2004.03.006

[22] Shahzad M et al. Computing the low dimension manifold in dissipative dynamical systems. The Nucleus. 2016;**53**(3):107-113

[23] Yablonsky GS, Constales D, Marin GB. Coincidences in chemical kinetics: Surprising news about simple reactions. Chemical Engineering Science. 2010;**65**(23):6065-6076. DOI: doi.org/10.1016/j.ces.2010.04.007

[24] Yablonsky GS, Constales D, Marin GB. Equilibrium relationships for non-equilibrium chemical dependencies. Chemical Engineering Science. 2011;**66**(1):111-114. DOI: doi.org/10.1016/j.ces.2010.10.014

[25] Yablonsky GS et al. Reciprocal relations between kinetic curves. EPL (Europhysics Letters). 2011;**93**(2):20004

[26] Constales D et al. Thermodynamic time-invariances: Theory of TAP pulse-response experiments. Chemical Engineering Science. 2011;**66**(20):4683-4689. DOI: doi.org/10.1016/j.ces.2011.06.033

[27] Yablonsky GS, Constales D, Marin GB. New types of complexity in chemical kinetics: intersections, coincidences, and special symmetrical relationships. In: Proceedings of the 240 Conference: Science's Great Challenges, Vol. 157; 12 December 2014; John Wiley & Sons, Inc.; 2014. DOI: 10.1002/9781118959602.ch6

[28] Muhammad Shahzad, 'Different Available Completion Routes in Complex Chemical Reactions'. International Conference on Mathematics in (bio) Chemical Kinetics and Engineering – MaCKiE2017 held on 25-27 May 2017 in Budapest, Hungary. http://static.akcongress.com/downloads/mackie/mackie2017-book-of-abstracts.pdf

Chapter 2

Mathematical Modeling and Simulation of Nonlinear Process in Enzyme Kinetics

Lakshmanan Rajendran, Mohan Chitra Devi,
Carlos Fernandez and Qiuming Peng

Additional information is available at the end of the chapter

http://dx.doi.org/10.5772/intechopen.70914

Abstract

A deep and analytical understanding of the enzyme kinetics has attracted a great attention of scientists from biology, medicine, chemistry, and pharmacy. Mathematical models of enzyme kinetics offer several advances for this deep and analytical understanding due to their in compensable potential in predicting kinetic processes and anticipating appropriate interventions when required. This chapter concerns mathematical modeling analysis and simulation of enzyme kinetics. Experimental data and available knowledge on enzyme mechanics are used in constituting a mathematical model. The models are either in the form of linear or nonlinear ordinary differential equations or partial differential equations. These equations are composed of kinetic parameters such as kinetic rate constants, initial rates, and concentrations of enzymes. The nonlinear nature of enzymatic reactions and a large number of parameters have caused major issues with regard to efficient simulation of those reactions. In this work, an enzymatic system that includes Michaelis-Menten and Ping Pong kinetics is modeled in the form of differential equations. These equations are solved numerically in which the system parameters are estimated. The numerical results are compared with the results from an existing work in literature.

Keywords: mathematical modeling, enzyme kinetics, chemical kinetics, nonlinear reaction-diffusion equation, amperometric, cyclic voltammetry, chronoamperometric

1. Introduction

Enzyme kinetics is a challenging research field nowadays incorporating modern applied mathematics into biotechnology, engineering science, and pharmacy. Moreover, in medical studies, scientists work on human metabolism to improve the capabilities of some metabolites or enzymes in metabolic pathways. In industrial applications, kinetics methods are also widely used to develop certain methods for improving functionality of some molecules in a cell. Many problems in theoretical and experimental biology/chemistry involve the solution of the steady-state reaction diffusion equation with nonlinear chemical kinetics. Such problems also arise in the formulation of substrate and product material balances for enzymes immobilized within particles [1, 2], in the description of substrate transport into microbial cells [3–5], in membrane transport, in the transfer of oxygen to respiring tissue [6, 7], and in the analysis of any artificial kidney system [8].

To impose the functionality of some molecules in a cell, a mathematical model of such metabolic systems must be constructed and simulated. Most of the dynamical systems can be approximated by various types of differential and integral equations involving finite number of variables and parameters. Thus, the future behavior of the system can be predicted if model kinetics parameters and initial states of the variables are available. In particular, ordinary and partial differential equations (ODEs and PDEs) are popular in modeling of the metabolic pathways or enzyme kinetics.

Releasing enzyme-substrate reactions under single-molecule kinetics was reported by Shlomi et al. [9]. An integral equation method with Michaelis-Menten kinetics to solve nonlinear diffusion problems in spherical coordinates was stated by Tosaka and Miyale [10]. Maalmi et al. [11] reported numerical and semianalytical solutions of nonlinear equations, which covered diffusivity, size, bulk concentration of reactant, binding constant of Michaelis-Menten kinetics, and site reactivity values. Merchant [12] stated the M-M decay reaction terms and the Gray-Scott scheme along with the semianalytical method to nonlinear reaction-diffusion systems. Indira and Rajendran [13] described a homotopy perturbation method to obtain substrate and product concentrations within the enzymatic layers. Removal of substrate from Michaelis-Menten kinetics governed the extravascular partition in which the analytical solution for the steady-state condition was investigated by Bucolo and Tripathi [14]. Dang Do and Greenfield [15] utilized the finite integral transform method to elucidate the problem based on the nonlinear reaction diffusion coupled with the chemical kinetics of a general shape solid. Chapwanya et al. [16] conveyed an epidemiological model with the Michaelis-Menten contact rate formulation to investigate variations in the enzyme kinetics with a simple susceptible infected recovered (SIR) model. Napper [17] proposed the Michaelis-Menten kinetics model to investigate the oxygen transport to heart tissue. Regalbuto et al. [18] presented an analytical methodology for obtaining solutions based on the maximum principle to nonlinear reaction-diffusion boundary value problems.

Rajendran and Saravanakumar [19] discussed mediated bioelectrocatalysis in order to build bioreactors, bio fuel cells, and biosensors.

Due to the difficulties in solving nonlinear differential equations in enzyme kinetics, some recent advanced analytical and numerical simulation techniques are used to solve the problems in chemical kinetics. Thus, in this review, all analytical and numerical works in enzyme kinetics are summarized.

2. Reaction diffusion systems

Reaction diffusion system is a mathematical model based on how the concentration of substances/products is disseminated over space changes under the influence of diffusion and a local chemical reaction. The substances are transformed into each other in local chemical reaction, whereas the substances are spread out over a surface in space in diffusion. Reaction-diffusion (RD) systems arise in many branches of physics, chemistry, biology, ecology, etc. Reviews of the theory and applications of reaction-diffusion systems can be found in books and numerous articles (see, for example [20–23]). These arise in a large variety of application areas, such as flow in porous media [24], heat conduction in plasma [25], combustion problems [26], liquid evaporation [27], and of more recent interest, image processing [28]. A great effort is being made in the development of the mathematical theory of nonlinear diffusion equations and to obtain exact solutions for special cases. Their significance not only relies on the huge number of their applications but also on the fact that they provide with a rather general class of linear and nonlinear differential operators. In mathematical analysis, it has shown to be a milestone for the development of applied, abstract, and numerical analysis as well as for algebra, geometry, and topology.

3. Nonlinear phenomena

The modern theory of the nonlinear reaction diffusion process is an important field in today's science. The nonlinear system and coherent structures represent an interdisciplinary area with many nonlinear applications in various fields. Those applications can be divided into six disciplines: chemistry (autocatalytic chemical and enzyme reactions), physics (nonlinear optics and electric circuits, plasmas and states of solid, condensed atomic gases, hydrodynamics, galaxy dynamics and cosmology, fluid dynamics, and celestial mechanics), general relativity, biology (biofuel cell, bioreactor and biosensor, atmosphere and oceans, and animal dispersal), random media, and modern telecommunications. A great variety of phenomena in physics, chemistry, or biology can be described by nonlinear ODE/PDEs and particularly by reaction-diffusion equations. For these reasons, the theory of the analytical solutions of the reaction-diffusion equations is considered.

In reaction diffusion systems, nonlinear phenomena play a crucial role in applied mathematics and chemistry. Exact (closed-form) solution of nonlinear reaction diffusion equations plays an important role in the proper understanding of qualitative features of many phenomena and processes in various areas of natural science. The main result obtained from reaction and diffusion systems is that nonlinear phenomena include diversity of stationary and spatio-temporary dissipative patterns, oscillations, different types of waves, excitability, biostability, etc. But it is difficult for us to obtain the exact solution for these problems. The investigation of exact solution of nonlinear equation is interesting and important. In general, this results in the need to solve linear and nonlinear reaction diffusion equations with complex boundary conditions. The enzyme kinetics in biochemical systems have usually been modeled by differential equations, which are based only on reaction without spatial dependence of the various concentrations. The dimensionless nonlinear reaction diffusion equations are described below:

$$\frac{\partial S}{\partial \tau} = \nabla^2 S - f(R, \tau, S, P) \tag{1}$$

$$\frac{\partial P}{\partial \tau} = \nabla^2 P + g(R, \tau, S, P) \tag{2}$$

where S and P represent the dimensionless concentrations of substrate and product, τ represents the dimensionless time, and R is the dimensionless radial co-ordinate of the particle. The first term on the right-hand side of the above equation accounts for active species (substrate or product) diffusion, whereas the second term $f(R,\tau,S,P)$ and $g(R,\tau,S,P)$ represents the homogeneous reaction term (nonlinear term), generally polynomial in the concentrations and time.

4. Common geometries and nonlinear reaction

Most commonly used electrodes/microelectrodes consist of a conducting metal/glassy carbon or semiconducting surface embedded in an insulating wall. When the conducting surface is a rectangle or disc of a few millimeters, this is known as a "planar" electrode. Diffusion to this surface is effectively planar (the effects of the edges are negligible), hence the nonlinear one-dimensional reaction diffusion equation is given by:

$$\frac{\partial [C]}{\partial t} = D\frac{\partial^2 [C]}{\partial x^2} + f([C]) \tag{3}$$

Two other electrode geometries where diffusion occurs in only one spatial dimension are the hemispherical and hemicylindrical electrodes. The nonlinear two-dimensional (hemispherical or spherical) reaction diffusion equation is:

$$\frac{\partial [C]}{\partial t} = D\left(\frac{\partial^2 [C]}{\partial x^2} + \frac{2}{r}\frac{\partial [C]}{\partial r}\right) + f([C]) \tag{4}$$

and for the latter is:

$$\frac{\partial [C]}{\partial t} = D\left(\frac{\partial^2 [C]}{\partial x^2} + \frac{1}{r}\frac{\partial [C]}{\partial r}\right) + f([C]) \tag{5}$$

The hemisphere can be achieved experimentally via a small drop of mercury positioned over a smaller conducting disc. A soft polymer, rubber, or other similar materials are usually employed to fabricate a hemicylinder. The electrodes are usually employed in theoretical studies due to the low dimensionality of the mass-transport equation. Additional terms such as diffusion and nonlinear reaction allow the equation to be solved analytically. Furthermore, the electrodes are not accurately or easily fabricated for practical geometries.

The corresponding nonlinear reaction-diffusion issues in enzyme kinetics are focused on the mathematical resolution. **Table 1** shows the response of particular electrodes with special emphasis on earlier theoretical works in the field.

Example 1: Michaelis-Menten kinetics and microcylinder electrodes

The model is written for an enzyme reaction to generate an electro-active product (e.g., hydrogen peroxide from an oxidase enzyme) that reacts at an immobilization matrix, which

Author	Reference	Experimental technique	Enzymatic scheme	Modeling method
		Analytical solutions		
G. Rahamathunissa et al.	Journal of theoretical and Computational Chemistry, 7(1)(2008)113–138	Amperometric	$S + C \xrightarrow{K_M} [SC] \rightarrow \left[PC'\right] \xrightarrow{k_c} P + C\,C' \xrightarrow{k_E'} C$	Danckwort's expression
R. Senthamarai et al.	Electrochemical Acta 53(2008)3566–3578	Chronoamperometric	$A + e \rightarrow B$ $B + Z \xrightarrow{k} A + product$	Analytical
G. Rahamathunissa L. Rajendran	Journal Mathematical Chemistry 44(2008)849–801	Amperometric	$E + S \overset{K_M}{\leftrightarrow} ES \xrightarrow{K_2} E + P$	Variation iteration method (VIM)
A. Meena et al.	Journal Mathematical Chemistry, 48(2010)179–186	Amperometric	$E + S \underset{K_{-1}}{\overset{K_1}{\rightarrow}} ES \rightarrow E + P$	He's variation iteration method
A. Eswari, L. Rajendran	Journal of Electroanalytical Chemistry 641(2010) 35–44	Amperometric	$S + E_1 \underset{K_2}{\overset{K_1}{\leftrightarrow}} [E_1 S] \xrightarrow{K_{cat}} P + E_2$	Homotopy perturbation method (HPM)
P. Manimozhi et al.	Sensors and Actuators B 147(2010)290–297	Amperometric	$E + S \underset{k_{-1}}{\overset{k_1}{\rightarrow}} ES \xrightarrow{k_c} E + P$ $ES + S \leftrightarrow ES_3$	Variational iteration and homotopy perturbation method (VIM & HPM)
S. Logambal, L. Rajendran	Electrochemical Acta 55(2010)5230–5238	Amperometric	$A + E_2 \xrightarrow{K_A} B + E_1$ $E_1 + S \xrightarrow{K_E} E_2 + P$	Homotopy perturbation method (HPM)
A. Meena, L. Rajendran	Journal of Electroanalytical Chemistry, 6411 (2010)50–59	Amperometric and Potentiometric	$E + S \leftrightarrow [ES] \rightarrow E + P$	Homotopy perturbation method (HPM)
S. Anitha, L. Rajendran	Journal of Physical Chemistry 114(2010)7030–7037	Amperometric	$B \xrightarrow{D_E} B + S \xrightarrow{K} A + Z \rightarrow A \rightarrow B$	Reduction of order method
P. Manimozhi, L. Rajendran	Journal of Electroanalytical Chemistry 647(2010) 87–92	Amperometric	$S + E \underset{K_\alpha}{\overset{K_a}{\leftrightarrow}} ES$ $ES \xrightarrow{K_{cat}} S' + E$	Analytical
A. Eswari, L. Rajendran	Journal of Electroanalytical Chemistry 648(2010) 36–46	Amperometric	$S + E_1 \underset{K_{-1}}{\overset{K_1}{\leftrightarrow}} [E_1 S] \rightarrow P + E_2$	Homotopy perturbation method (HPM)
A. Eswari, L. Rajendran	Russian Journal of Electroanalytical Chemistry 47(2011)195–204	Cyclic voltammetry	$E\,A + e \leftrightarrow B$ $C\,B \xrightarrow{k_1}$ Products	Laplace Transformation
A. Eswari, L. Rajendran	Russian Journal of Electroanalytical Chemistry 47(2011)205–212	Cyclic voltammetry	$E\,A + e \leftrightarrow B$ $C_2\,B + B \xrightarrow{k_1}$ Products	Homotopy perturbation method (HPM)

Author	Reference	Experimental technique	Enzymatic scheme	Modeling method
A. Eswari, L. Rajendran	Journal of Electroanalytical Chemistry 651(2011) 173–184	Chronoamperometric	$O + ne^- \leftrightarrow R$ $R + Z \xrightarrow{k} O + \text{Products}$	Homotopy perturbation method (HPM)
G. Rahamathunissa et al.	Journal of Mathematical Chemistry 9(2011)457–474	Chronoamperometric	$S + E \overset{k_M}{\leftrightarrow} ES \xrightarrow{k_2} E + P$	VIM
S. Logambal, L. Rajendran	Journal of Membrane Sciences 373(2011)20–28	Amperometric	$E_{OX} + S \underset{k_M}{\overset{k_{-1}}{\leftrightarrow}} ES \xrightarrow{k_2} E_{red} + P$ $E_{red} + O_2 \xrightarrow{k_3} E_{OX} + H_2O_2$	Homotopy perturbation method (HPM)
S. Anitha et al.	Electrochimica Acta 56(2011)3345–3352	Amperometric	$S + E_1 \overset{K_M}{\leftrightarrow} [E_1S] \xrightarrow{k_{cat}} P + E_2A \rightarrow B$	Homotopy perturbation method (HPM)
K. Indra, L. Rajendran	Electrochimica Acta 56(2011)6411–6419	Chronoamperometric	$S_1 + O_2 \xrightarrow{PPO} P_2 + H_2O\ V_1$ $P_2 + 2e^- + 2H^+ \underset{k_r}{\overset{k_0}{\leftrightarrow}} S_2\ E^0$ $S_2 + \frac{1}{2}O_2 \xrightarrow{PPO} P_2 + H_2O\ V_2$	Homotopy perturbation method (HPM)
S. Thiagarajan et al.	Journal of Mathematical Chemistry DOI: 10.1007/s10919-011-9854-z	Chronoamperometric	$S + M_{ox} \overset{k_M}{\leftrightarrow} SM_{ox} \xrightarrow{k_{cat}} P + M_{red}$	Homotopy perturbation method (HPM)
M. Uma Maheswari, L. Rajendran	Journal of Mathematical Chemistry DOI: 10.1007/s10919-011-9853-0	Chronoamperometric	$E + S \overset{K_1}{\leftrightarrow}{}_{k-1} ES \xrightarrow{k_2} E + P$	Homotopy perturbation method (HPM)
P. Rijiravanich et al.	Electroanalytical Chemistry 589(2006)249	Amperometric	$O_2 + 2catechol \rightarrow 2o - quinone + 2H_2O$ $o - quinone + 2H^+ + 2e^- \rightarrow catechol$	Theory and experiment
A. Eswari, L. Rajendran	Journal of Electroanalytical Chemistry 660(2011) 200–208	Amperometric	$O_2 + 2catechol \rightarrow 2o - quinone + 2H_2O$ $o - quinone + 2H^+ + 2e^- \rightarrow catechol$	VIM
G. Varatharajan, L. Rajendran	Applied Mathematics 2(2011)1140–1147	Amperometric	$S + E \underset{k-1}{\overset{K_1}{\leftrightarrow}} C \xrightarrow{k_{cat}} P + E$ $E \xrightarrow{k_3} E_i$	Homotopy perturbation method (HPM)
K. Venugopal et al.	Journal of Biomedical Science and Engineering 4 (2011)631–641	Chronoamperometric	$O_2 + 2catechol \rightarrow 2o - quinone + 2H_2O$ $o - quinone + 2H^+ + 2e^- \rightarrow catechol$	Homotopy perturbation method (HPM)
K. Indra, L. Rajendran	Journal of Mathematical Chemistry DOI: 10.1007/s10910-011-9968-3	Chronoamperometric	$A \leftrightarrow B + C$ $B \pm e^- \rightarrow products$	Homotopy perturbation method (HPM)
V. Margret Ponrani, L. Rajendran	Journal of Mathematical Chemistry DOI: 10.1007/s10910-011-9973-6	Amperometric	$G + E \underset{K_1}{\overset{k_{-1}}{\leftrightarrow}} X \underset{k_2}{\overset{k_{-2}}{\rightarrow}} F + E$	Homotopy perturbation method (HPM)

Author	Reference	Experimental technique	Enzymatic scheme	Modeling method
S. Sevukaperumal et al.	Applied Mathematics 3(2012)373–381	Chronoamperometric	$Glucose + O_2 \overset{Glucoseoxidase}{\longrightarrow} gluconicacid + H_2O_2$ $H_2O_2 \overset{Catalase}{\longrightarrow} H_2O_2 + \frac{1}{2}O_2$	Homotopy analysis method (HPM)
		Numerical solution		
R. Baronas et al.	Biosensors and Bioelectronics 19(2004)915–922	Amperometric	$S \rightarrow P \overset{E}{\rightarrow} S$	Finite-difference technique
R. Baronas	Electrochimica Acta 240(2017)399–407	Amperometric biosensor	$S + E \underset{k_{-1}}{\overset{k_1}{\leftrightarrow}} ES \rightarrow P + E$ $S \overset{E}{\rightarrow} P$	Numerical simulation and analytical solution
V. Ašerisa et al.	Journal of Electroanalytical Chemistry 685(2012) 63–71	Amperometricparallel substrates conversion	$S_1 \overset{E_1}{\rightarrow} \frac{1}{2} P_1$ $S_1 + S_2 \overset{E_2}{\rightarrow} P_2$	Digital simulation-finite-difference technique
V. Flexer et al.	Bioelectrochemistry 74(2008)201–209	Cyclic voltammetry	$S + E_{OX} \underset{k_{-1}}{\overset{k_1}{\leftrightarrow}} ES \overset{k_{cat}}{\rightarrow} P + E_{red}$	Numerical simulation
R. Baronas et al.	Chemometrics and Intelligent Laboratory Systems 126(2013)108–116	Amperometric	$E + S_i k_{1_i} \leftrightarrow ES_i \overset{k_{2_i}}{\rightarrow} E + P_i, i = 1, \ldots, k$	Numerical
R. Baronas	Nonlinear Analysis: Modeling and Control 9(3) (2004)203–218	Amperometric	$S \overset{E}{\rightarrow} P$	Digital simulation-finite-difference technique.
R. Baronas et al.	Sensors 12(2012)9146–9160	Amperometric	$E_{OX} + S \overset{k_1}{\rightarrow} E_{red} + P$ $E_{red} \overset{k_2}{\rightarrow} E_{OX} + n_e e^-$	Finite-difference
R. Baronas et al.	J. Mathematical Chemistry 32 (2)(2002)225–237	Amperometric	$S \overset{E}{\rightarrow} P$	Numerical simulation
R. Baronas et al.	Mathematical Modeling of Biosensors, Springer Series on chemical sensors and biosensors (2009)	Amperometric	All enzyme reactions	Analytical and numerical methods
L. Rajendran	Biosensor: Modeling and Simulation of Diffusion-Limited Process, Chemical Sensors: Simulation and Modeling, GhenadiiKorotcenkov (Ed.), Electrochemical Sensors, Vol. 5, Momentum Press, LLC, New York (2013)	Amperometric	All enzyme reactions	Analytical, HPM&HAM, VIM,ADM, etc.

Table 1. Contributions to the theoretical modeling of enzymatic electrodes.

is metallically conducting sites/particles. The reaction within the film under the Michaelis-Menten kinetics may be written as follows:

$$S + E_1 \underset{k_{-1}}{\overset{k_1}{\Leftrightarrow}} [E_1 S] \xrightarrow{k_{cat}} P + E_2 \tag{6}$$

The consumption rate of S is given by $k_1 c_S c_E - k_{-1} c_{ES}$, where c_i denotes the concentration of species i. The rate is equivalent to $(k_{cat}/K_M)\ c_S c_E$, where K_M is the Michaelis constant, defined as $K_M = (k_{-1} + k_{cat})/k_1$. The consumption rate of S in the film is compensated by diffusion. If the solution is stirred uniformly, so that S is constantly supplied to the film, the mass balance for S can be written in cylindrical coordinates:

$$\frac{D_S}{r}\frac{d}{dr}\left(r\frac{dc_S}{dr}\right) - \frac{k_{cat} c_E c_S}{c_S + K_M} = 0 \tag{7}$$

where c_S is the concentration profile of substrate, c_E is the concentration profile of enzyme, D_S is its diffusion coefficient, and K_M is the Michaelis constant. The rate of consumption will be $v(r) = k\ c_H$, where k is the rate constant for the hydrogen peroxide reaction and c_H is the peroxide concentration. Then, the equation of continuum for hydrogen peroxide is generally expressed in the steady-state by

$$\frac{D_H}{r}\frac{d}{dr}\left(r\frac{dc_H}{dr}\right) + \frac{k_{cat} c_E c_S}{c_S + K_M} - v(r) = 0 \tag{8}$$

At the electrode surface (r_0) and at the film surface (r_1), the boundary conditions are [29]:

$$\begin{aligned} r = r_0 &: \quad \frac{dc_S}{dr} = 0, \quad c_H = 0 \\ r = r_1 &: \quad c_S = c_S^*, \quad c_H = 0 \end{aligned} \tag{9}$$

where c_S^* is the bulk concentration of S scaled by the partition coefficient of the film. The current is provided by the consumption rate at each site. Thus, the total current at an electrode of length L is expressed by [29]

$$I/nF = 2\pi L \int_{r_0}^{r_1} rv\, dr \tag{10}$$

The analytical results of the problem are discussed by Eswari and Rajendran [30].

Example 2: enzyme catalysis reaction

The reactions without spatial dependence on various concentrations have modeled the enzyme kinetics in biochemical systems. Nonlinear systems of ordinary differential equations are solely based on that. Michaelis and Menten were pioneers in explaining the enzyme reaction model. In addition, they also reported the free enzyme binding to the reactant, which produced an enzyme-reactant complex. Eq. (11) illustrates the Michaelis-Menten kinetics, in which the enzyme-substrate complex is formed after the enzyme is combined with the substrate.

$$E + S \underset{k_{-1}}{\overset{k_1}{\leftrightarrow}} ES \overset{k2}{\rightarrow} E + P \tag{11}$$

As can be seen from Eq. (11), the product P is released by the binding of substrate S with enzyme E. The product released is not reversible; however, the substrate binding is reversible. The reactants' concentrations in Eq. (11) are represented by the following letters:

$$s = [S], \quad e = [E], \quad c = [SE], \quad p = [P] \tag{12}$$

The law of mass action leads to the system of following nonlinear reaction equations [31],

$$\frac{ds}{dt} = -k_1 es + k_{-1} c \tag{13a}$$

$$\frac{de}{dt} = -k_1 es + (k_{-1} + k_2) c \tag{13b}$$

$$\frac{dc}{dt} = k_1 es - (k_{-1} + k_2) c \tag{13c}$$

$$\frac{dp}{dt} = k_2 c \tag{13d}$$

where k_1 is the forward rate of ES complex formation and k_{-1} is the backward rate constant. The above problem is discussed theoretically by Meena et al. [32].

Example 3: Michaelis-Menten mechanism for co-substrate and substrate

Figure 1 illustrates Michaelis-Menten reaction kinetics scheme for co-substrate and substrate. Limoges et al. [33] reported for a redox enzymatic homogenous system along with one-dimensional mass transport equation a concise discussion and derivation.

When the enzyme is being solubilized, the electrochemical signal that is produced during the reaction is governed by the following set of nonlinear partial differential equations.

$$\frac{\partial [Q]}{\partial t} = D_P \frac{\partial^2 [Q]}{\partial x^2} - \frac{C_E^0}{\frac{1}{k_1 [S]} + \frac{1}{k_{1,2}} + \frac{1}{k_{2,2}} + \frac{1}{k_2 [Q]}} \tag{14}$$

$$\frac{\partial [S]}{\partial t} = D_S \frac{\partial^2 [S]}{\partial x^2} - \frac{C_E^0}{\frac{1}{k_1 [S]} + \frac{1}{k_{1,2}} + \frac{1}{k_{2,2}} + \frac{1}{k_2 [Q]}} \tag{15}$$

where D_P, D_S are the diffusion coefficients of co-substrate and substrate, respectively; Q, S are the concentrations of co-substrate and substrate, respectively; x is the distance from the

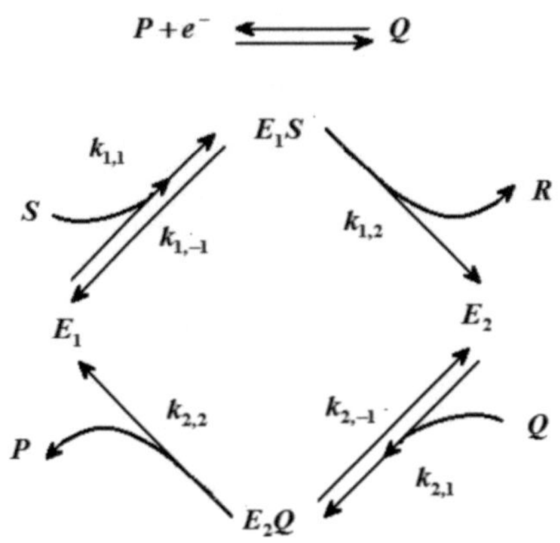

Figure 1. Reaction scheme for substrate and co-substrate.

electrode surface; C_S^0 is the bulk concentration of substrate; C_E^0 is the total concentration of enzyme; k_1, $k_{2,2}$, and k_2 are the reaction rate constants; and t is the time. The initial and boundary conditions for Eqs. (14) and (15) are given by:

$$t = 0, x \geq 0, \text{and}\, x = \infty, x \geq 0, [Q] = 0, [S] = C_S^0 \tag{16}$$

$$x = 0, t \geq 0 : [Q] = \frac{C_P^0}{1 + \exp\left[\frac{F}{RT}\left(E - E_{PQ}^0\right)\right]}, \frac{\partial[S]}{\partial x} = 0 \tag{17}$$

$$x = \infty, \partial[Q]/\partial x = 0 \tag{18}$$

The analytical expressions corresponding to the concentration of co-substrate for steady and nonsteady state conditions have been obtained by solving the above nonlinear equation using a new approach to homotopy perturbation method (HPM). Analytical expressions of the plateau current are also presented for steady and nonsteady state conditions:

$$i = FSD_P\left(\frac{\partial[Q]}{\partial x}\right)_{x=0} \tag{19}$$

where E is the electrode potential, E_{PQ}^0 is the standard potential of the P/Q couple, F is the Faraday constant, and S is the surface area of the electrode. The above problem is discussed theoretically by Rasi et al. [34].

5. Analytical solutions

To study many of the physical phenomena, the exact solutions of nonlinear partial or ordinary differential equations play an important role. In order to understand the mechanism of complicated dynamical processes and physical phenomena modeled by nonlinear differential equations, the existence of approximate analytical and exact solutions is very important. In addition, nonlinear differential equations can also assist to investigate the stability of these solutions as well as checking the simulation analysis. Nonlinear partial differential equations govern a significant variety of phenomena including physical, chemical, and biological. The development of techniques aimed at exact solutions of nonlinear differential equations with nonsteady and steady state [35] has been one of the most exciting advances of nonlinear science and theoretical physics/chemistry. An important role in nonlinear science is played by exact solutions of differential equations. Furthermore, this can be especially observed in nonlinear physical chemistry science. This can be attributed to the provision of physical information as well as more insight into the physical aspects of the problem, which could lead to further applications. Over the past few decades, different methods have been reported to solve analytical solutions such as Tanh-sech [36], extended tanh [37], Jacobi elliptic function expansion [39], hyperbolic function [38], F-expansion [40], and the First integral [41]. To solve different types of nonlinear systems of PDEs, the sine-cosine method [42] has been employed. A variety of powerful analytical methods such as homotopy perturbation method [43–45], homotopy analysis method [46, 47], Adomian decomposition method [48, 49], wavelet transform method [50], etc. are applied to solve the nonlinear problems (e.g., Eqs. (8) and (13)–(15)) in chemical kinetics [51].

6. Numerical solutions

Many differential equations cannot be solved analytically. For practical purpose, however, such as in physical engineering sciences, a numerical approximation to the solution is often sufficient. The numerical method is mainly to solve complex problem physically or geometrically. It is also used to validate the experimental results. Some of the nonlinear equations in chemical kinetics were solved using numerical methods [52–56].

7. Summary

Most mathematical models of enzyme kinetics are based on reaction diffusion equations or rate equations containing nonlinear terms related to the kinetics of the enzyme reaction. Powerful and accurate analytical (HPM, HAM, ADM, etc.) and numerical mathematical methods have been employed for their resolution under steady and nonsteady state conditions. The theoretical results provide very useful insight into the effects on the performance of the thickness and structure of the enzymatic film, the loading of the different species, the diffusivity of the mediator, etc. Also, the theoretical modeling and simulation of these systems enable us to characterize the enzymatic reactions (i.e., rate constant, turnover rate, and Michaelis-Menten constants).

In spite of the above-mentioned benefits, there are only limited theoretical studies addressing kinetics of enzyme reaction and most of them include a number of simplifying assumptions mainly related to the mass and charge transport inside and outside the biocatalyst film, the enzymatic kinetic scheme, and the electrode morphology. Experimental validation of proposed models is even more seldom. Therefore, more effort in the future research is needed in this direction in order to develop more detailed models and accurate simulations that can assist the rational development and optimization of enzyme electrodes.

Author details

Lakshmanan Rajendran[1*], Mohan Chitra Devi[1], Carlos Fernandez[2] and Qiuming Peng[3]

*Address all correspondence to: raj_sms@rediffmail.com

1 Department of Mathematics, Sethu Institute of Technology, Pulloor, Kariapatti, India

2 School of Pharmacy and Life Sciences, Robert Gordon University, UK

3 State Key Laboratory of Metastable Materials Science and Technology, Yanshan University, Qinhuangdao, China

References

[1] Aris R. The Mathematical Theory of Diffusion and Reaction in Permeable Catalysts. Vol. 1. Clarendon: Oxford; 1975

[2] Engasser JM, Horvath C. Effect of internal diffusion in heterogeneous enzyme systems: Evolution of true kinetic parameters and substrate diffusity. Journal of Theoretical Biology. 1973;**42**:137-155

[3] Cheviollotte P. Relation between the reaction cytochrom oxidase-oxygen and oxygen uptake in cells in vivo: The role of diffusion. Journal of Theoretical Biology. 1973;**39**:277-295

[4] McElwain DLS. A re-examination of oxygen diffusion in a spherical cell with Michaelis-Menten oxygen uptake kinetics. Theoretical Biology. 1978;**71**:205-263

[5] Lin SH. Oxygen diffusion in a spherical shell with nonlinear oxygen uptake kinetics. Journal of Theoretical Biology. 1976;**60**:449-457

[6] Ho SP, Kostin MD. Diffusion with irreversible chemical reaction in heterogeneous media: Application to oxygen transport in respiring tissue. Journal of Theoretical Biology. 1997;**64**:237-251

[7] Pope AS. Diffusion in tissue slices with metabolism obeying Michaelis-Menten kinetics. Journal of Theoretical Biology. 1979;**80**:325-332

[8] Lim SH. A modified model for predicting the performance of a compact artificial kidney. Journal of Theoretical Biology. 1972;**77**:441-451

[9] Shlomi R, Michael U, Joseph K. Role of substrate unbinding in Michaelis-Menten enzymatic reactions. Proceedings of the National Academy Sciences. 2012;**111**:4391-4396

[10] Tosaka N, Miyale S. Analysis of a nonlinear diffusion problem with Michaelis-Menten kinetics by an integral equation method. Bulletin of Mathematical Biology. 1982;**44**(6):841-849

[11] Maalmi M, Strieder W, Varma A. Ligand diffusion and receptor mediated internalization: Michaelis-Menten kinetics. Journal of Chemical Engineering Science. 2001;**56**(19):5606-5616

[12] Merchant TR. Cubic autocatalysis with Michaelis-Menten kinetics: Semi-analytical solutions for the reaction-diffusion cell. Journal of Chemical Engineering Science. 2004;**59**(16): 3433-3440

[13] Indira K, Rajendran L. Analytical expression of the concentration of substrates and product in phenol-polyphenol oxidase system immobilized in laponite hydrogels; Michaelis-Menten formalism in homogeneous medium. Electrochimica Acta. 2011;**56**(18):6411-6419

[14] Bucolo J, Tripathi K. Steady-state analysis of a two-compartment barrier-limited capillary-tissue model with Michaelis-Menten saturation kinetics. Bulletin of Mathematical Biology. 1980;**42**(5):691-700

[15] Dang Do D, Greenfield F. A finite integral transform technique for solving the diffusion-reaction equation with Michaelis-Menten kinetics. Journal of Mathematical Biosciences. 1981;**54**(1–2):31-47

[16] Chapwanya M, Lubuma S, Mickens E. From enzyme kinetics to epidemiological models with Michaelis-Menten contact rate design of nonstandard finite difference schemes. Journal of Computers and Mathematics with Applications. 2012;**64**(3):201-213

[17] Napper A, Schubert RW. Michaelis–Menten kinetics as a modelling assumption in a model of oxygen transport in heart Proceedings of the First Southern Biomedical Engineering Conference. 1981;201-204

[18] Regalbuto C, Strieder W, Varma A. Approximate solutions for nonlinear diffusion-reaction equations from the maximum principle. Journal of Chemical Engineering Science. 1988;**43**(3):513-518

[19] Rajendran L, Saravanakumar K. Analytical expression of transient and steady-state catalytic current of mediated bioelectrocatalysis. Journal of Electrochimica Acta. 2014;**147**:678-687

[20] Nicolis G, Prigogine I. Self-Organization in Non-equilibrium Systems. New York: Wiley; 1977

[21] Mikhailov AS. Foundations of Synergetics. Berlin: Springer-Verlag; 1990

[22] Kerner BS, Osipov VV. A New Approach to Problems of Self-Organization and Turbulence. Autosolitons. Dordrecht: Kluwer; 1994

[23] Lubashevskii A, Gafiychuk VV. The projection dynamics of highly dissipative system. Physical Review E. 1994;**50**(1):171

[24] Vázquez JL. The Porous Medium Equation. Mathematical Theory. Oxford Mathematical Monographs. Oxford: Oxford University Press; 2006

[25] Bertsch M. Asymptotic behavior of solutions of a nonlinear diffusion equation. SIAM Journal on Applied Mathematics. 1982;**42**(1):66

[26] Aronson DG, Weinberger HF. Nonlinear diffusion in population genetics, combustion and nerve pulse propagation. In: Goldstain JA, editor. Partial Differential Equations and Related Topics, Lecture Notes in Mathematics. Berlin/New York: Springer-Verlag; 1975; pp. 446

[27] Okrasiski W, Parra MI, Cuadros F. Modeling evaporation using a nonlinear diffusion equation. Journal of Mathematical Chemistry. 2001;**30**(2):195

[28] Mikula K, Ramarosy N. Semi-implicit finite volume scheme for solving nonlinear diffusion equations in image processing. Numerische Mathematik. 2001;**89**(3):561-590

[29] Somasundrum M, Aoki K. The steady-state current at microcylinder electrodes modified by enzymes immobilized in conducting or non-conducting material. Journal of Electroanalytical Chemistry. 2002;**530**(1–2):40-46

[30] Eswari A, Rajendran L. Analytical solution of steady state current an enzyme modified micro cylinder electrodes. Journal of Electroanalytical Chemistry. 2010;**648**:36-46

[31] Murray JD. Mathematical Biology. Vol. 175. Berlin: Springer Verlag; 1989

[32] Meena A, Eswari A, Rajendran L. Mathematical modelling of enzyme kinetics reaction mechanisms and analytical solutions of nonlinear reaction equations. Journal of Mathematical Chemistry. 2010;**48**:179-118

[33] Limoges B, Moiroux J, Savéant J-M. Kinetic control by the substrate and/or the co-substrate in electrochemically monitored redox enzymatic homogeneous systems catalytic responses in cyclic voltammetry. Journal of Electroanalytical Chemistry. 2002;**521**:1-7

[34] Rasi M, Rajendran L, Subbiah A. Analytical expression of transient current-potential for redox enzymatic homogenous system. Sensors and Actuators B Chemical. 2015;**B208**:128-136

[35] Alquran M, Al-Khaled K, Ananbeh H. New soliton solutions for systems of nonlinear evolution equations by the rational sine-cosine method. Studies in Mathematical Sciences. 2011;**3**(1):1-9

[36] Malfliet W. Solitary wave solutions of nonlinear wave equations. American Journal of Physics. 1992;**60**(7):650-654

[37] El-Wakil SA, Abdou MA. New exact travelling wave solutions using modified extended tanh-function method. Chaos Solitons Fractals. 2007;**31**(4):840-852

[38] Xia TC, Li B, Zhang HQ. New explicit and exact solutions for the Nizhnik-Novikov-Vesselov equation. Applied Mathematics E-Notes. 2001;**1**:139-142

[39] Inc M, Ergut M. Periodic wave solutions for the generalized shallow water wave equation by the improved Jacobi elliptic function method. Applied Mathematics E-Notes. 2005;**5**: 89-96

[40] Zhang S. The periodic wave solutions for the 2+1-dimensional Konopelchenko Dubrovsky equations. Chaos Solitons Fractals. 2006;**30**:1213-1220

[41] Feng ZS. The first integer method to study the Burgers-Kortewegde Vries equation. Journal of Physics A: Mathematical and General. 2002;**35**(2):343-349

[42] Mitchell AR, Griffiths DF. The Finite Difference Method in Partial Differential Equations. John Wiley & Sons: Chichester-New York-Brisbane-Toronto; 1980

[43] Shanthi D, Ananthaswamy V, Rajendran L. Analysis of nonlinear reaction-diffusion processes with Michaelis-Menten kinetics by a new homotopy perturbation method. Natural Science. 2013;**5**(9):1034-1046

[44] Saranya J, Rajendran L, Wang L, Fernandez C. A new mathematical modelling using homotopy perturbation method to solve nonlinear equations in enzymatic glucose fuel cells. Chemical Physics Letters. 2016;**662**:317-326

[45] Meena A, Rajendran L. Mathematical modeling of amperometric and potentiometric biosensors and system of nonlinear equation homotopy perturbation approach. Journal of Electroanalytical Chemistry. 2010;**644**:50-59

[46] Angel Joy R, Meena A, Loghambal S, Rajendran L. A two-parameter mathematical model for immobilized enzymes and homotopy analysis method. Natural Science. 2011;**37**: 556-565

[47] Kirthiga M, Chitra Devi M, Meena A, Rajendran L. Mathematical modelling and kinetics of micro channel reactor. Applied and Computational Mathematics. 2016;**56**:234-246

[48] Sivasankari MK, Rajendran L. Analytical expression of the concentration of species and effectiveness factors in porous catalysts using the Adomian decomposition method. Kinetics and Catalysis. 2013;**54**(1):95-105

[49] Renuga Devi M, Sevukaperumal S, Rajendran L. Nonlinear reaction diffusion equation with Michaelis-Menten Kinetics and adomian decomposition method. Applied Mathematics. 2015;**51**:21-32

[50] Mahalakshmi M, Hariharan G, Kannan K. The wavelet methods to linear and nonlinear reaction-diffusion model arising in mathematical chemistry. Journal of Mathematical Chemistry. 2013;**51**(9):2361-2385

[51] Rajendran L. Biosensor: Modeling and simulation of diffusion-limited process, chemical sensors: simulation and modeling. Ghenadiikorotcenkov ed. Electrochemical sensors. 2013;**5**

[52] Bard AJ, Faulkner LR. Electrochemical Methods: Fundamentals and Applications, 2nd ed., John Wiley & Sons, INC: New York; 2001; pp. 864

[53] Britz D. Digital Simulation in Electrochemistry. Springer-Verlag: Berlin Heidelberg; 1988

[54] Compton RG, Banks CE. Understanding Voltammetry. 2nd ed. Imperial College Press: London WC2H 9HE; 2010; pp. 444

[55] Bieniasz LK. Towards computational electrochemistry: A Kineticist's perspective. Modern Aspects of Electrochemistry-Springer. 2002;**35**:135-195

[56] Baronas R et al. Mathematical Modelling of Biosensors. Springer Series on Chemical Sensors and Biosensors - Springer Netherlands; 2009

Chapter 3

Autoignition and Chemical-Kinetic Mechanisms of Homogeneous Charge Compression Ignition Combustion for the Fuels with Various Autoignition Reactivity

Dongwon Jung

Additional information is available at the end of the chapter

http://dx.doi.org/10.5772/intechopen.70541

Abstract

This work demonstrates the autoignition and chemical-kinetic mechanisms of homogeneous charge compression ignition (HCCI) combustion for the fuels with various autoignition reactivity. This is done for four fuels: methane, dimethyl ether (DME), iso-octane and n-heptane. Methane and iso-octane are selected as the single-stage ignition fuel, and DME and n-heptane are selected as the two-stage ignition fuel. As a tool for understanding the characteristics of autoignition and combustion process in HCCI engine, a zero-dimensional single-zone engine model of 'CHEMKIN' in Chemkin-Pro was used. The complete compression and expansion strokes were modeled using an engine with a connecting-rod length to crank-radius ratio of 3.5 and a compression ratio of 13. A detailed chemical-kinetic mechanism for methane and DME is Mech_56.54 (113 species and 710 reactions). For iso-octane and n-heptane, a detailed chemical-kinetic mechanism from Lawrence Livermore National Laboratory (1034 species and 4236 reactions) is used. The results show that methane and iso-octane exhibit only the main heat release, 'high-temperature heat release (HTHR)' by high-temperature reactions (HTR). In contrast, both DME and n-heptane exhibit the first heat release 'low-temperature heat release (LTHR)' associated with low-temperature reactions (LTR) before HTHR.

Keywords: homogeneous charge compression ignition, autoignition, chemical-kinetic mechanisms, fuel autoignition reactivity, low-temperature reaction, intermediate-temperature reaction, high-temperature reaction

1. Introduction

1.1. Background

Since their introduction around a century ago, internal combustion engines have played a key role in shaping of the modern world [1]. Because of their simplicity, ruggedness and high power/weight ratio, internal combustion engine has found wide application in transportation [2]. Though there are technologies that could theoretically provide more environmentally sound alternatives, internal combustion engines, such as fuel cells and electric vehicles, practically, cost, efficiency and power density issues, will prevent them displacing internal combustion engines in the near future. However, in recent decades, serious concerns have been raised with regard to the environmental impact of emissions arising from operation of internal combustion engines. Eventually, concerns about climate change lead to ever-stricter fuel-economy legislations [2-4]. In addition, concerns about the world's finite oil reserves result in heavy taxation of road transport, mainly via on duty on fuel [5]. These two factors have led to massive pressure on vehicle manufacturers to research, develop and produce ever cleaner and more fuel-efficient vehicles. Ultimately, all legislations for emissions from vehicles are targeted to improve technologies to the point where an affordable, practical zero emissions vehicle (ZEV) with outstanding performance becomes a reality [6]. Even though there are many types of real ZEVs, operated by fuel cells that consume hydrogen generated from water by electricity produced from renewable sources, it is very unlikely that the resulting vehicles could even come close to meeting any of the other criteria listed above in the short and medium terms [1, 2]. For this reason, the bulk of vehicle research and development resources are still being applied to the internal combustion engines to increase their efficiency.

1.2. Homogeneous charge compression ignition (HCCI) engine

The purpose of internal combustion engines is the production of mechanical power from the chemical energy contained in the fuel. This chemical energy is released by burning or oxidizing the fuel inside the engine. The fuel-air mixture before combustion and burned products after combustion are the actual working fluids. The work transfers, which provide the desired power output, occur directly between these working fluids and the mechanical components of the engine.

As **Figure 1** shows, there are three main types of internal combustion engines:

- Spark ignition (SI) engine
- Compression ignition (CI) engine
- Homogeneous charge compression ignition (HCCI) engine

http://dx.doi.org/10.5772/intechopen.70541

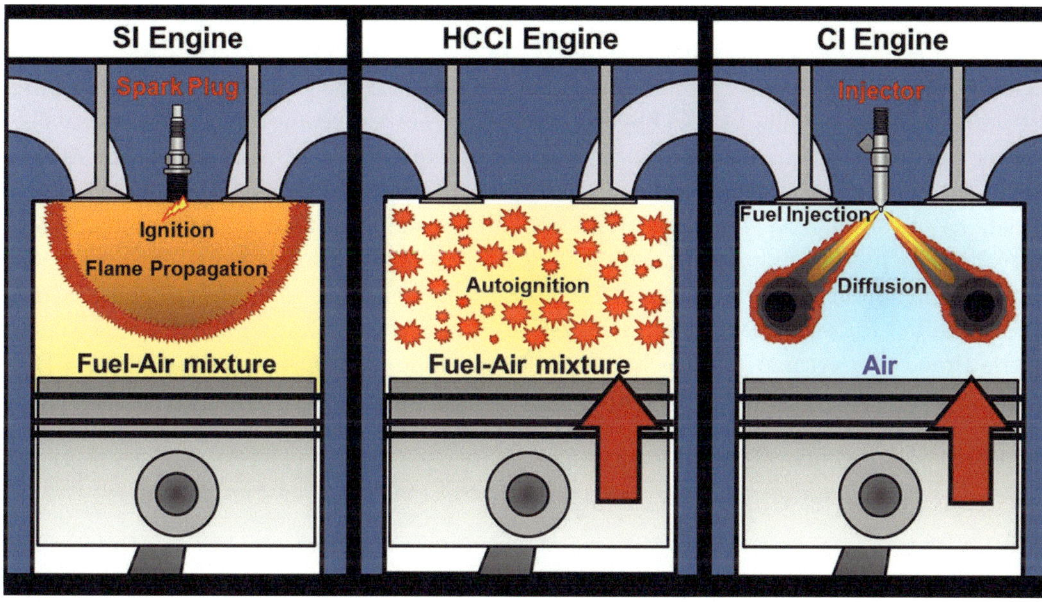

Figure 1. Illustration of the combustion characteristics for spark ignition (SI) engine (left), compression ignition (CI) engine (right) and homogeneous charge compression ignition (HCCI) engine (middle).

In a SI engine, the fuel and air are mixed together in the intake system, inducted through the intake valve into the cylinder. Then, the fuel-air mixture is compressed towards the end of the compression stroke, and the combustion is initiated by a spark discharge at the spark plug. By spark discharge, an inflammation is occurred, and then a turbulent combustion developed fully through the premixed fuel-air mixture until it reaches the combustion chamber walls, and then extinguishes [7]. For CI engine, fuel is injected by the fuel-injection system in the engine cylinder towards the end of the compression stroke, just before the desired start of combustion. The liquid fuel is usually injected at high velocity through small nozzles in the injector tip, and atomized into small drops while penetrating into the engine in-cylinder. Then, with the high-temperature high-pressure in-cylinder air, the fuel is vaporized and mixed. When the in-cylinder air temperature and pressure are increased above the fuel's ignition point, the ignition of portions of the already-mixed fuel and air occurs spontaneously after a delay period of a few milliseconds. The consequent compression of the unburned portion of the charge shortens the delay before ignition for the fuel and air which has mixed within combustible limits, which then burns rapidly [8]. Fundamentally, the HCCI combustion is a controlled autoignition of the homogeneous mixture through compression by piston [9]. To a degree, the HCCI engine is able to combine the best feature of SI engine and CI engine [10]. Similar to the SI engine, the fuel and air are mixed together and inducted to obtain a homogenous mixture, which can eliminate fuel-rich diffusion combustion, and can thus dramatically reduce the particulate matter (PM) that is the main problem of CI combustion. With the ignition process, similar to that of CI combustion, HCCI combustion is achieved through the autoignition of the fuel-air homogeneous mixture around the top dead centre (TDC) as it is compressed via the piston, which can lead to very low nitrogen oxides (NO_x) by reducing a high-temperature flames when compared to that of SI combustion. Furthermore, the unthrottled operation of HCCI engines with relatively high compression ratio is possible at a very low fuel/air equivalence ratio (φ) and a high rate of external exhaust gas recirculation (EGR) without misfire, thus yielding a high thermal efficiency with a very low cycle-to-cycle variations of combustion. Therefore, the HCCI combustion is an attractive technology that can ostensibly provide engine efficiencies comparable to that of diesel engine, and engine-out emissions comparable to or less than that of SI engine with a three-way catalyst. These advantages have led to considerable interest in HCCI in recent years and to substantial research efforts aimed at overcoming the technical challenges to its widespread implementation [11]. The technical challenges are briefly summarized as follows:

- Combustion-phasing control
- Excessive heat-release rate (HRR) at high loads
- Narrow operating range

The successful operation of an HCCI engine depends on using mechanical means to control both the autoignition and the combustion processes. The heat-release rate (HRR) from HCCI combustion depends not only on the unique reaction chemistry of the fuel but also on the thermal conditions that the in-cylinder charge mixture goes through during compression by piston. To enable to control the start of combustion as well as the overall combustion rate for HCCI combustion, it is critically important to have a resolute understanding of the interaction between the chemical-kinetic mechanisms of the fuel-air mixture and the history of in-cylinder temperature and pressure during the compression and expansion strokes.

2. Chemical-kinetics modelling setup for numerical calculation

2.1. Zero-dimensional single-zone engine model

A zero-dimensional single-zone engine model (referred to here as 'single-zone model') of CHEMKIN [12] in Chemkin-Pro [13] was used for this work. Using an engine with a connecting-rod length to crank-radius ratio of 3.5 and a compression ratio of 13, the complete compression and expansion strokes (i.e. from compression bottom dead centre (BDC) to expansion BDC) were modeled according to the standard slider-crank relationship [14]. The crevices and boundary layers were not included. The numerical calculations were conducted under following assumptions:

- The in-cylinder charge is treated as a single lumped mass with uniform mixture composition and thermodynamic properties (homogenous in-cylinder charge).
- The in-cylinder charge is compressed and expanded adiabatically (adiabatic change).
- All species present in the in-cylinder charge are considered as the ideal gas (following the ideal gas law).

- The total energy of the in-cylinder charge remains constant (following the first law of thermodynamics).
- For the in-cylinder charge, the total mass of products equal to the total mass of reactants (following the law of conservation of mass).

These are great oversimplification of a real engine in which the fuel-air mixture will never be completely mixed and there will always be residuals from the previous cycle (mixture inhomogeneity). In addition, there are at least four causes for temperature inhomogeneity: (1) heat transfer from the in-cylinder charge to the cylinder wall, (2) presence of hot residuals from the previous cycle as a result of incomplete mixing, (3) dynamic flow effects during the intake stroke and (4) vaporization of the fuel, especially if injected directly into the cylinder. Because both mixture and temperature inhomogeneities for the in-cylinder charge will significantly affect the heat-release rate, the burn duration, the peak in-cylinder charge pressure, the peak combustion temperature and the amount of emissions, the single-zone model cannot accurately predict these values. Nonetheless, the single-zone model can provide useful results in at least two ways. First, the single-zone model has an advantage for predicting the autoignition timing with a reasonable accuracy because the autoignition timing is dominated by the autoignition reactions of the hottest zone in the core of in-cylinder charge. It can be thought as representing the close-to-adiabatic core in the experiment because the single-zone model is adiabatic. This indicates that the changes in the autoignition timing with EGR addition and boosting, and the amount of initial in-cylinder charge temperature at BDC required to compensate for these changes in the autoignition timing are realistic values. Second, the single-zone model is a useful tool for investigating certain fundamental aspects of HCCI combustion, since eliminating the complexities of mixture-temperature inhomogeneities, heat transfer, blow-by, and crevices and boundary layers simplifies the analysis and allows cause-and-effect relationships to be more easily identified. This means that it allows the effects of the bulk-gas (gases not in crevices or boundary layers) chemical-kinetics and thermodynamics to be isolated in order to understand how they alone influence the autoignition and the combustion process.

2.2. Fuel selection

Since HCCI engine has the capability of operating with a variety of fuels, HCCI operation has been demonstrated for various fuels that have autoignition reactivity spanning a wide range. Although each fuel exhibits different autoignition reactivity even for the same experiment conditions, the autoignition characteristics of fuels can be broadly divided into two types: those with single-stage ignition fuel and those with two-stage ignition fuel which exhibits the first heat-release 'low-temperature heat release (LTHR)' associated with cool-flame chemistry before the main heat-release 'HTHR'. Many factors ultimately affect the choice of fuel, but each fuel-type has advantages for HCCI engines, respectively. A brief summary for the advantages of each fuel-type follows:

- Advantages of single-stage ignition fuel for HCCI engine
  - The use of high compressions ratio is allowed, which leads to high thermal efficiency.

- The ignition timing is much less sensitive to changes in speed and load than that of two-stage ignition fuel, which indicates that significantly less compensation will be required to maintain optimal ignition timing over the required load and speed range.
- Advantages of two-stage ignition fuel for HCCI engine
  - Because the amount of LTHR produced by a two-stage ignition fuel increases with the local fuel/air equivalence ratio (φ), fuel stratification can be used for controlling combustion phasing.
  - This local φ-dependence of the LTHR also provides a means for reducing the peak HRR.

For this work, methane and iso-octane are selected as the single-stage ignition fuel, and DME and n-heptane are selected as the two-stage ignition fuel. Methane and DME are classified as a gaseous fuel. On the other hand, iso-octane and n-heptane are classified as a liquid fuel. **Figure 2** presents the chemical formula and the illustration of chemical structure, and **Table 1** lists the properties for these four test fuels [15, 16].

2.3. Chemical-kinetic mechanism for test fuels

An overall reaction includes very complex and sophisticated reactions that cannot be analyzed without a proposed chemical-kinetic mechanism, a series of steps that a reaction takes before reaching the final products. The chemical-kinetic mechanism is step-by-step descriptions of what happens on a molecular level in chemical reactions. Each step of the reaction mechanism for the overall reaction is known an elementary reaction. The term elementary reaction is used to describe a moment in the reaction when one or more molecules change geometry or perturbed by the addition or omission of another interacting molecule. For methane and DME, a detailed chemical-kinetic mechanism (Mech_56.54; 113 species and 710 reactions) by Burke et al. [17] was used, which has been developed to be capable of predicting the combustion of both methane and DME in common combustion environments such as compression ignition engines and gas turbines. For iso-octane and n-heptane, a detailed

	Methane	DME	n-Heptane	iso-Octane
Chemical formula	CH_4	CH_3OCH_3	C_7H_{16}	C_8H_{18}
(H): Hydrogen (C): Carbon (O): Oxygen	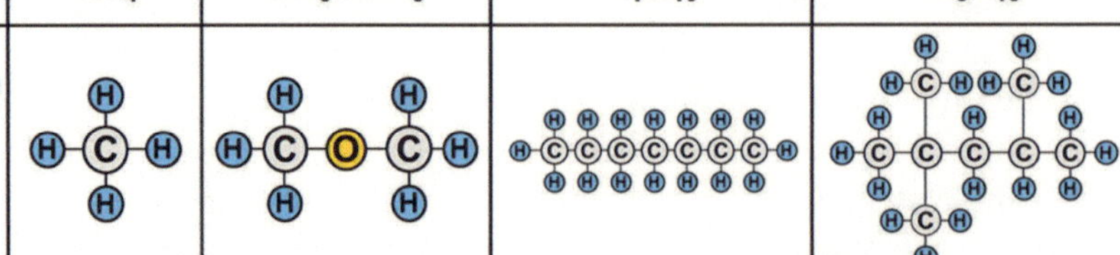			
Mechanism	Mech_56.54 [37]		PRF: iso-Octane/n-Heptane Mixtures [38]	
Species	113		1034	
Reaction	710		4236	

Figure 2. Chemical formula and illustration of chemical structure with reaction mechanism for methane, DME, n-heptane and iso-octane.

Property (unit)	Methane	DME	n-Heptane	iso-Octane
Boiling point (°C)	−161.5	−25.1	98.4	99.2
Liquid density (g/cm^3@20°C)	—	0.67	0.68	0.6878
Relative gas density (air = 1)	0.55	1.6	3.46	3.9
Vapor pressure (MPa)	—	0.61@25°C	0.0046@20°C	0.0051@20°C
Ignition temperature (°C)	650	235	285	417
Lower heat value (MJ/kg)	49.0	28.8	44.57	44.31

Table 1. Properties of DME [15], methane [15], n-heptane [16] and iso-octane [16].

chemical-kinetic mechanism from Lawrence Livermore National Laboratory (LLNL; 1034 species and 4236 reactions) [18] was used, which has been developed for the oxidation of primary reference fuels (PRFs), iso-octane and n-heptane, for gasoline. This mechanism was developed by combining the iso-octane [19] and n-heptane [20] mechanisms.

2.4. Comparison of autoignition delay times

Autoignition, the spontaneous ignition of a fuel and oxidizer mixture in the absence of any external ignition source, occurs when slow thermal reactions initially have a large chain branching component sufficiently to maintain and accelerate oxidation. The increasing radical concentration leads to the increase in reaction rate build on themselves, and eventually result in an ignition through a rapid explosive rise in radical concentration, oxidation rate and temperature. Most of these reactions typically release heat, and eventually increasing the temperature and pressure of the system, and at the same time, their rate is also strongly dependent on pressure, temperature and charge composition. These characteristics cause a complicated interaction of negative and positive feedback loops that determine when the ignition will happen. In fact, autoignition is very sensitive to details of chain branching and chain terminating in the initial reactions, and hence depends sensitively on the chemical structure of the fuel.

The autoignition reactivity of the fuel is a very important parameter, impacting the design and the potential high-load performance of HCCI engines. The accurate prediction of autoignition times and their dependence on pressure, temperature and composition is essential for advanced engine technologies, such as HCCI, where the ignition event is timed by chemical kinetics. An autoignition delay time (τ) of fuels is one of the crucial indicators to present the extent of fuel autoignition reactivity for the combustion optimization of internal combustion engines, especially for HCCI engines. The autoignition delay time is defined as the time interval required for the fuel-air mixture to spontaneously ignite at some prescribed conditions. The rapid compression machine (RCM) and shock tube are two of the most widely used facilities for the studies of ignition delay time. RCM gives a direct way of measuring the ignition delay time by simulating the process of adiabatic compression and ignition. While the shock tube is applied to study autoignition characteristics of gas mixtures at a higher temperature and pressure than those of RCM, RCM is used to study the autoignition characteristics of test fuels in the temperature range of low to intermediate, compared with shock tubes. To understand the reaction activity for the test fuels, the autoignition delay times are examined by conducting the numerical calculation using 'Closed Homogeneous Batch Reactor' in CHMEKIN-Pro with the chemical-kinetic mechanisms explained in the section describing the chemical-kinetic mechanism for test fuels. The initial pressure was set as 3.0 MPa to represent the maximum in-cylinder pressure for motored operation of engine modeled for this work. In addition, the initial fuel/air equivalence ratio (ϕ_o) was set as 0.5 since HCCI engines are generally operated with lean in-cylinder charge mixture. **Figure 3** compares the results of autoignition delay times between test fuels. As can be observed, methane does not exhibit any of the low-temperature reaction (LTR) or negative temperature coefficient (NTC) behavior typical of larger paraffinic fuels such as n-heptane. (The term 'negative temperature coefficient' is used to denote the temperature regime where the rate of fuel consumption decreases with increasing temperature, rather than increases as in all other regimes.) This indicates that methane is very resistant to autoignition and correspondingly has a very high octane number (=120). In contrast, for DME, the highest fuel autoignition reactivity (i.e. the shortest autoignition time) is observed until the initial temperature of 1170 K. In addition, DME displays NTC

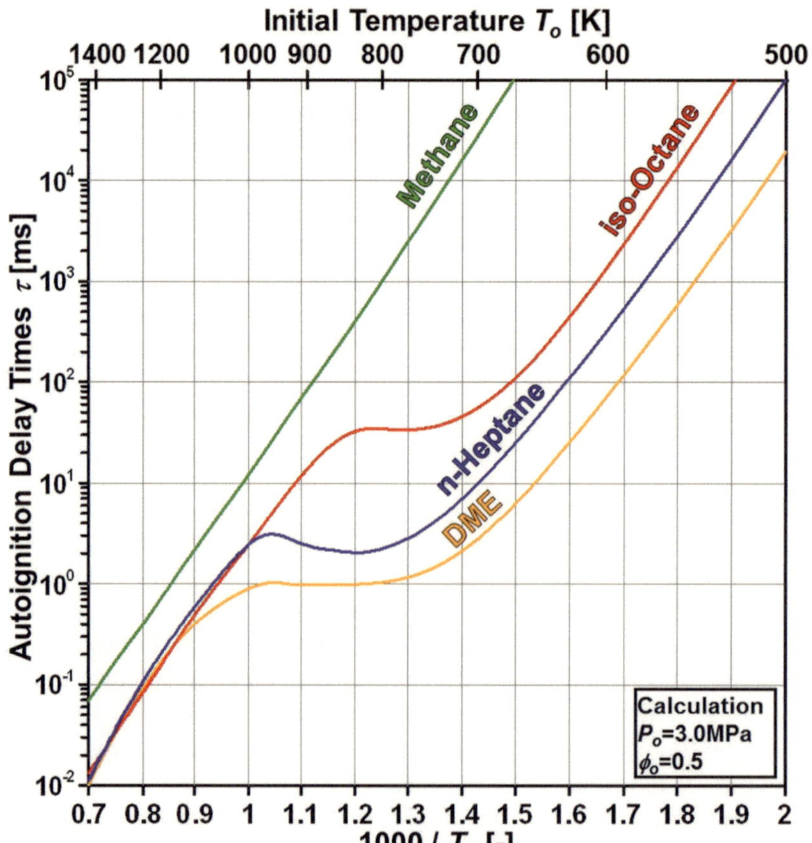

Figure 3. Comparison of autoignition delay times for methane, DME, n-heptane and iso-octane.

http://dx.doi.org/10.5772/intechopen.70541

behavior where the autoignition delay times increase with increasing initial temperature. Similar to that of DME, the autoignition delay time for n-heptane also shows NTC behavior with relatively high-fuel autoignition reactivity due to very low octane number (=0). For iso-octane, surprisingly, NTC behavior is observed despite of high octane number (=100) even for relatively longer than autoignition time delay of DME and n-heptane. As shown in Ref. [21], which systematically investigates autoignition properties of iso-octane at conditions relevant to practical combustion devices using RCM, iso-octane can exhibit NTC region under conditions of elevated initial pressure.

3. Chemistry of HCCI combustion

As discussed above, reactants in HCCI combustion begin at room temperature and are steadily heated during the compression stroke by piston. As the reactant temperature increases, the specific elementary reactions that contribute to fuel consumption in general and chain branching and autoignition in particular also change. The reactants pass through three distinct temperature ranges, each with its own unique chain branching reaction pathways that contribute to the eventual autoignition. With reference to **Figure 4** as an example of HCCI combustion, this section explains the chemical reactions that play a role in the process, which are classified as low-temperature reactions (LTR), intermediate-temperature reactions (ITR) and high-temperature reactions (HTR).

3.1. Low-temperature reactions

Virtually, no significant reaction takes place until the reactant temperature reach about 550 K. As the reactant heats up during the compression stroke, chemistry becomes increasingly active at temperatures above 600 K. At these conditions, fuel dissociation is described by the following low-temperature mechanism [22].

$RH + O_2$	$\Rightarrow$	$R\bullet + HO_2$	initiation	Re.(1)
$R\bullet + O_2$	$\Leftrightarrow$	$RO_2\bullet$	first O_2 addition	Re.(2)
$RO_2\bullet$	$\Rightarrow$	$\bullet R'OOH$	internal H-atom abstraction	Re.(3)
$\bullet R'OOH$	$\Rightarrow$	$R'O + OH\bullet$	chain propagation	Re.(4)
$\bullet R'OOH + O_2$	$\Leftrightarrow$	$HO_2R'O_2\bullet$	second O_2 addition	Re.(5)
$HO_2R'O_2\bullet + RH$	$\Rightarrow$	$HO_2R'O_2H + R\bullet$	external H-atom abstraction	Re.(6)
$HO_2R'O_2H$	$\Rightarrow$	$HO_2R'O\bullet + OH\bullet$	chain branching	Re.(7)
$HO_2R'O\bullet$	$\Rightarrow$	$OR'O + OH\bullet$	chain propagation	Re.(8)
$HO_2R'O_2\bullet$	$\Rightarrow$	$HO_2R''O_2H$	internal H-atom abstraction	Re.(9)
$HO_2R''O_2H\bullet$	$\Rightarrow$	$HO_2R''O + OH\bullet$	chain propagation	Re.(10)
$HO_2R''O$	$\Rightarrow$	$OR''O\bullet + OH\bullet$	chain branching	Re.(11)

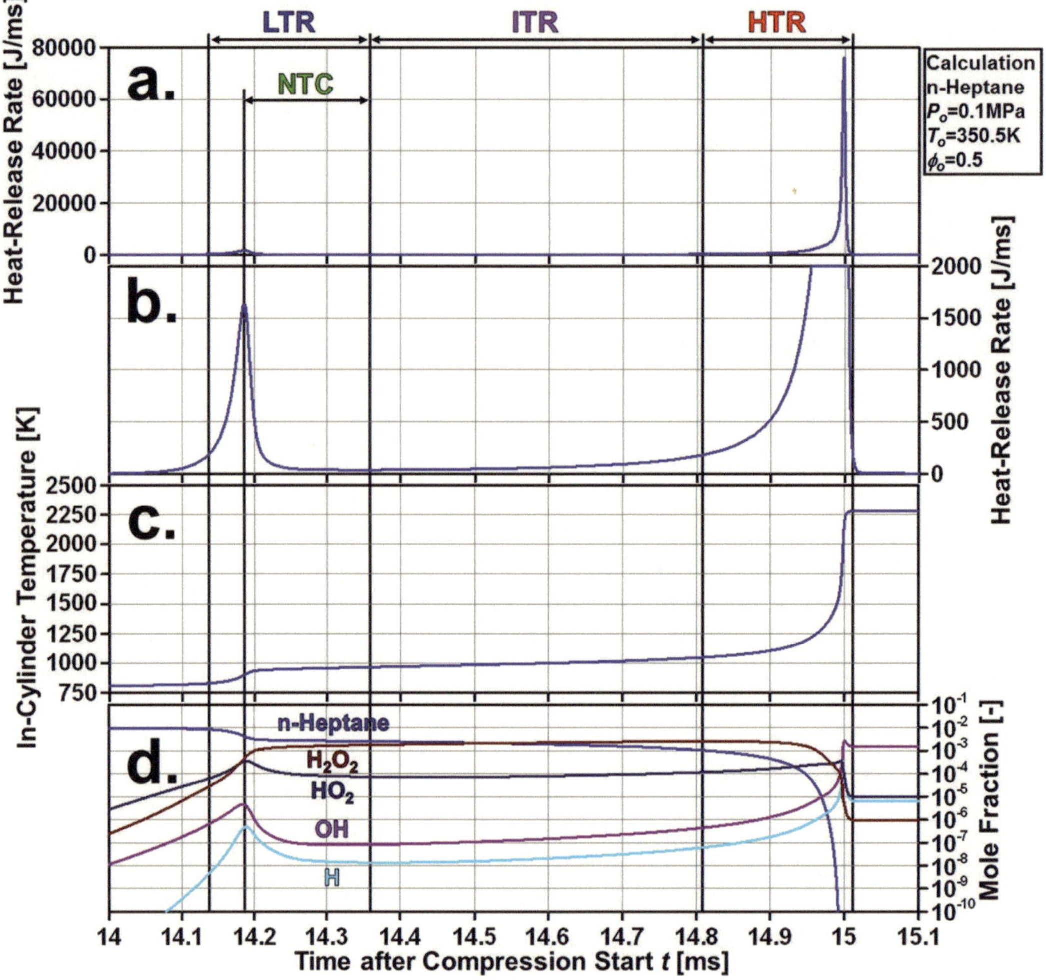

Figure 4. An example of HCCI combustion for (a) overall heat-release rate, (b) magnified view of heat-release rate, (c) in-cylinder temperature and (d) mole fraction.

In the initiation step, a hydrocarbon (RH) reacts with oxygen (O_2) to make a hydrocarbon radical (R•), which reacts with oxygen to make a peroxy radical (RO_2•). (Radical species are denoted by the '•' symbol next to the character.) Next, an internal hydrogen-atom abstraction takes place (i.e. the abstraction of a hydrogen atom from the molecule itself). Following the internal abstraction, the radical •R'OOH reacts internally to eliminate (eject) OH and forms a compound without free valences (unpaired electrons) such as an aldehyde or ketone (Reaction (4)). The mechanism continues with a second O_2 addition to the peroxy radical initially formed (Reaction (5); [23]). After a few steps, keto-hydroperoxide ($HO_2R''O$) is formed (Reaction (10)). Keto-hydroperoxide decomposes at around 800 K, producing further hydroxyl (OH•) that consumes the fuel (Reaction (11)). In the hydrogen abstraction reaction (Reaction (3)), the molecule isomerizes by 'reaching around' and abstracting a hydrogen atom from somewhere on the hydrocarbon chain. Straight chain molecules such as n-heptane are long enough for flexible internal abstraction of hydrogen (Reaction (1)). In addition to this, H atoms in n-heptane are bound to 'secondary sites' (the -CH_2- backbone), which makes them easier to abstract H atoms in primary sites, where the hydrogen is attached to the end of a chain (the -CH_3 group). Iso-octane is actually a short pentane chain with three methyl groups attached to the chain. The short chain has difficulty 'reaching around' to abstract a hydrogen atom and furthermore, most of the H atoms in iso-octane are primary, thus harder to abstract. This flexibility and abstraction theory explains the higher reactivity and lower octane number of n-heptane (octane number = 0) with respect to iso-octane (octane number = 100). The theory further explains the high octane number of methane (octane number = 120) where no internal abstraction is possible. The mechanism from Reaction (1) to Reaction (11) listed above also explains the observation of so-called 'two-stage ignition', also called 'negative temperature coefficient (NTC)' zone. At low temperature, the oxygen addition (Reactions (2) and (5)) leads to a product '**P**' that then undergoes reactions that lead to chain branching (Reactions (7) and (11)). These chain branching reactions lead to a rapid increase in the temperature of the mixture. As the temperature increases, the NTC zone is reached where the newly formed product '**P**' can now either continue towards chain branching or decompose beak to the reactants (i.e. reverse reaction, see the bi-directional arrow on Reactions (2) and (5)). The increase in the reverse rate results in a lower concentration of products '**P**' which in turn leads to a reduction of chain branching, causing a reduction in the rate of temperature increase; the ignition delay is prolonged. As a consequence, one observes what is called 'two-stage ignition'. At low temperatures, the reactions are proceeding at a slow, but observable rate. Starting at temperature below the NTC zone, the energy release by these reactions slowly increases the temperature. With this increased temperature, the reaction rates increase, the temperature is increasing faster and faster. This is the 'first stage' of ignition. The temperature increase until the NTC zone is reached. At this temperature, the concentration of '**P**' decreases, and thus the rate of increase in temperature slows down, but is never zero. With time, the slowly increasing temperature reaches a point where low concentration of product '**P**' is more than compensated by the increased chain branching reaction rate and then, the system explodes: this is the 'second stage' of ignition. Surprisingly, if one starts the system in the NTC zone, the concentration of '**P**' is extremely low and the ignition delay can be longer than if one stared the system at a temperature below the NTC zone. This is why it is called 'negative temperature coefficient (NTC)' zone.

3.2. Intermediate-temperature reactions

As the temperature increases above about 850 K, where the equilibria of Reactions (2) and (5) have effectively extinguished the low-temperature chain branching pathways, the next reaction sequences involve consumption of fuel (RH), primarily by hydrogen (H) atom abstraction by OH and hydroperoxyl (HO_2), and the temperature increases gradually, accompanied by a steady increase in the level of hydrogen peroxide (H_2O_2), as shown in **Figure 4d**. This new set of chemical reactions contributing to the increase in the level of H_2O_2 with the increase of temperature is called 'intermediate-temperature reactions (ITR)' and is described by the following main intermediate-temperature mechanism [24].

$H_2O_2 + M + 216kJ$	⇒	$OH + OH + M$	Re.(12)
$OH + CH_2O$	⇒	$HCO + H_2O + 122kJ$	Re.(13)
$HCO + O_2$	⇒	$CO + HO_2 + 138kJ$	Re.(14)
$HO_2 + HO_2$	⇒	$H_2O_2 + O_2 + 168kJ$	Re.(15)

These four reactions, which are also called 'H_2O_2 loop reactions', show similar activation energies and reaction rates. The rate constants of Reactions from (13) to (15) are significantly larger than that of Reaction (12), and their activation energies are very small. These suggest that Reactions from (12) to (15) compose a reaction loop in which the rate-determining process is Reaction (12), as schematically shown in **Figure 5**. On the assumption that 100% of OH, HCO and HO_2 generated by 'H_2O_2 loop reactions' are consumed by the succeeding reactions, the overall reaction is to be Reaction (16).

$2CH_2O + O_2$	⇒	$2H_2O + 2CO + 472kJ$	Re.(16)

This is a reaction to release a considerable amount of heat from CH_2O without consuming H_2O_2. In addition, following reactions support 'H_2O_2 loop reactions' by supplying the key species of formaldehyde, formyl (HCO) and OH.

$C_2H_3 + O_2$	⇒	$CH_2O + HCO + 359kJ$	Re.(17)
$CH_2CHO + O_2$	⇒	$CH_2O + CO + OH + 212kJ$	Re.(18)
$CH_3 + O$	⇒	$CH_2O + H + 293kJ$	Re.(19)

The sub intermediate-temperature mechanism, Reactions from (17) to (20), also participates in the process.

$H + O_2 + M$	⇒	$HO_2 + M + 202kJ$	Re.(20)
$H_2O_2 + OH$	⇒	$H_2O + HO_2 + 130kJ$	Re.(21)
$HO_2 + OH$	⇒	$H_2O + O_2 + 297kJ$	Re.(22)
$HO_2 + CH_2O$	⇒	$H_2O_2 + HCO - 7kJ$	Re.(23)

Reaction (20) enhances 'H_2O_2 loop reactions' by supplying HO_2 using H generated mainly by Reaction (19). In the real process, Reaction (23) and Reactions from (17) to (19) contribute to additional H_2O_2.

3.3. High-temperature reactions

The heat-release rate by intermediate-temperature reactions grows steadily, until at about 1000 K, four important events occur. The H_2O_2, which has been relatively stable due to the strength of its O–O bond and the correspondingly large value of the activation energy of its decomposition reaction, begins to decompose at ever-increasing rates by following reaction.

$H_2O_2 + M$	⇒	OH• + OH• + M	chain branching	Re.(24)

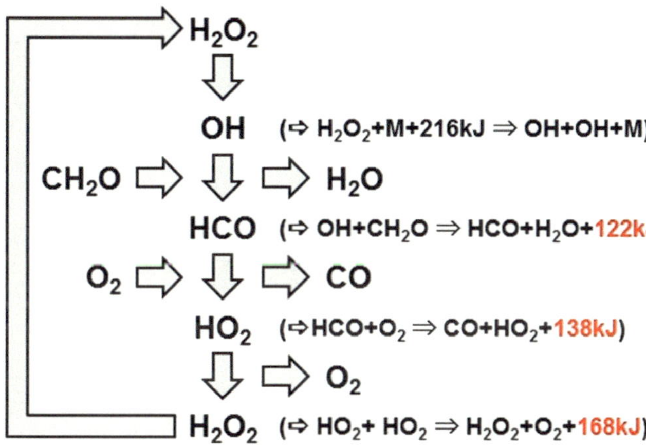

Figure 5. Schematic of H_2O_2 reaction loop [24].

This decomposition causes the concentration of OH• to grow very quickly. The importance of Reaction (24) is clearly seen in **Figure 4d** for mole fraction, where the concentration of H_2O_2 decreases rapidly during HCCI combustion as OH radicals are being formed, increasing the temperature of the reacting mixture and setting in motion as effective chain branching sequence. As a result, the fuel is very rapidly consumed by reacting with this sudden source of OH•, and the temperature increases very rapidly, due to the production of significant amounts of water by reaction of 'RH + OH ⇒ R• + H_2O', further accelerating the rate of H_2O_2 decomposition. All of these events occurring together create an autoignition event. The Reaction (24) sequence proceeds until the temperature has increased sufficiently that the high-temperature chain branching sequence take over, controlled by H• + O_2 ⇒ O• + OH• which dominates the remainder of the overall HCCI combustion process. The decomposition of H_2O_2 (Reaction (24)) 'triggers' ignition in HCCI combustion. This reaction has a critical temperature for ignition that is also a function of the pressure of the reactive system. H_2O_2 decomposition can be written, ignoring for the moment all other reactions of H_2O_2, by the simple differential equation

$$\frac{d[H_2O_2]}{dt} = -k_5[H_2O_2][M] \quad (1)$$

where M is the molar concentration and k_5 is the rate of Reaction (24). This equation can be rearranged to define a characteristic decomposition time (α).

$$\alpha = [H_2O_2]/(d[H_2O_2]/dt) = 1/(k_5[M]) \quad (2)$$

The rate expression for this reaction is

$$k_5 = 1.2 \times 10^{17} \times \exp(-45500/RT) \quad (3)$$

so the characteristic decomposition time α becomes

$$\alpha = 8.3 \times 10^{-18} \times \exp(+22750/T) \times [M]^{-1} \tag{4}$$

As the temperature increases, τ becomes smaller and also decreases with increasing total concentration $[M]$ or, equivalently, with increasing pressure at constant temperature. This also means that, as pressure increases, the critical temperature for ignition decreases gradually. Usually, the autoignition occurs at temperature between 1050 and 1100 K. The consistency of this temperature is a recognized feature of HCCI combustion. Not surprisingly, this temperature is comparable to the ignition temperature that is observed during engine knock in SI engines [25].

4. Comparison of combustion characteristic between test fuels in HCCI engine

This section will compare the combustion characteristics of the test fuels in HCCI engine. As discussed in conjunction with **Figure 3**, each respective test fuel shows different autoignition delay times even for the same initial condition due to its fuel autoignition reactivity. Because of this, we should expect quiet different combustion phasing for each fuel depending on the resistance to autoignition under the constant initial condition. The combustion phasing is a critical parameter impacting the thermal efficiency of HCCI engine. If the combustion is too advanced, knocking combustion occurs easily, thus quickly increasing to the risk for engine damage and NO_x emissions. On the other hand, excessive combustion-phasing retard leads to unacceptable coefficient of variation (COV) of HCCI combustion with partial-burn and/or misfire cycles. To facilitate comparison of the combustion characteristics in HCCI engine, the initial temperature is adjusted in the numerical simulation to set the 50% burn point (CA50) at 0degATDC (i.e. TDC). Effectively, the reported combustion phasing refers to CA50 for the main combustion event, starting at the crank angle of minimum heat-release rate between LTHR and HTHR. Presenting the data referring to the main combustion event alone is considered more relevant from the standpoint of quantifying the onset of the main combustion event.

With the effects of fuel autoignition reactivity isolated, **Figure 6** compares (a) in-cylinder temperature, (b) heat-release rate, (c) magnified view of heat-release rate and (d) accumulated heat release for the test fuels. The required initial temperature (T_o) to maintain CA50 = 0degATDC is 559.5, 301.5, 464.d and 350.5 K for methane, DME, iso-octane and n-heptane, respectively. If the fuel has high resistance to autoignition (i.e. methane and iso-octane), high in-cylinder charge temperature is required during the compression stroke in order to ensure autoignition. As can be seen, methane and iso-octane both require relatively high T_o, which typically has a negative influence on the peak load that can be obtained. This happens because these two fuels exhibit single-stage ignition at this calculation condition. For a given initial pressure (P_o), higher T_o causes high peak combustion temperature, as shown in **Figure 6a**, so excessive NO_x can become the load-limiting factor. Furthermore, as **Figure 6b** shows, the increase of peak combustion temperature contributes to the increase

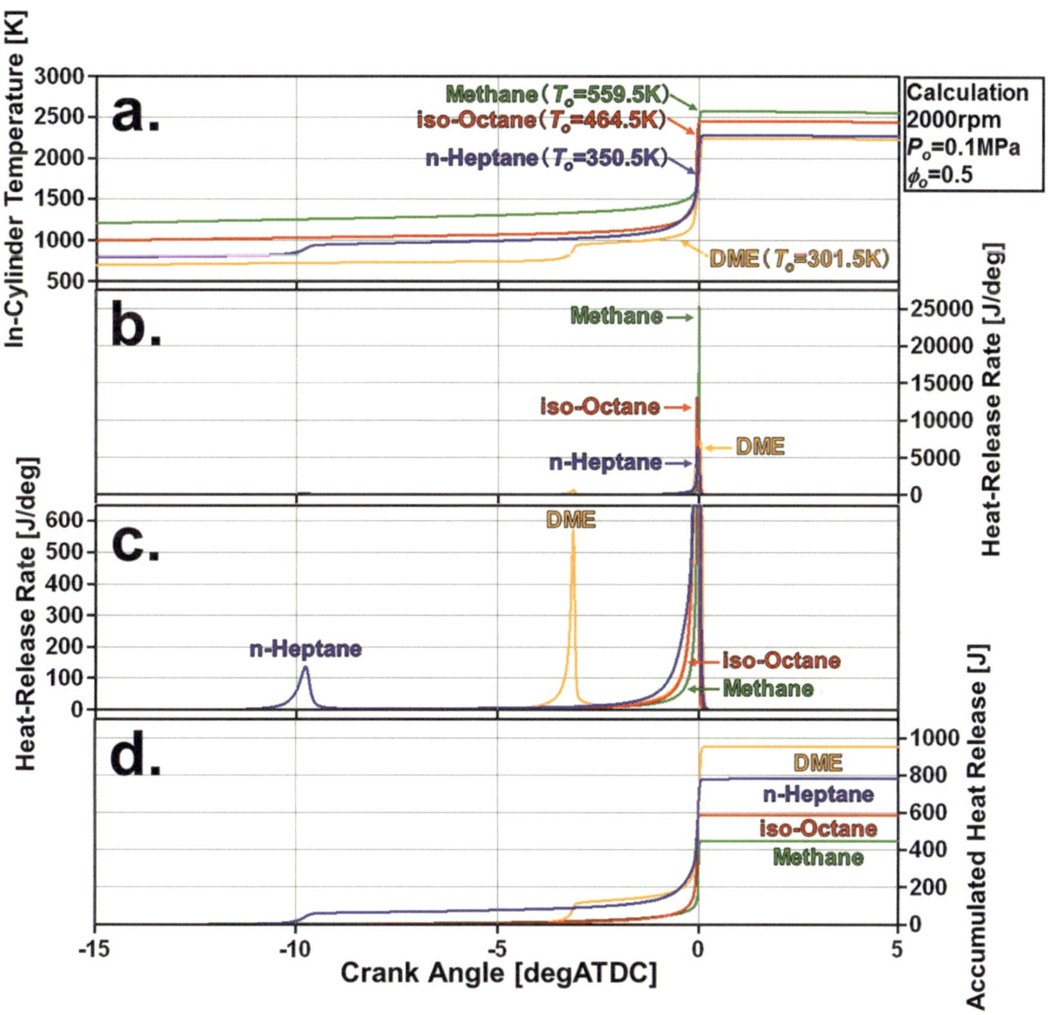

Figure 6. Comparison of (a) in-cylinder temperature, (b) heat-release rate, (c) magnified view of heat-release rate and (d) accumulated heat release for methane, DME, iso-octane and n-heptane.

of maximum heat-release rate. In addition, the higher T_o leads to lower in-cylinder charge density and thus the smaller amount of fuel when φ_o is constant. This can be explained by the data in **Figure 6d** for the accumulated heat release, which shows the lower accumulated heat release for the higher T_o. DME shows the highest accumulated heat release and methane the lowest. This is an important aspect that potentially can help to increase the higher power output. On the other hand, DME and n-heptane exhibit two-stage ignition with LTHR for this calculation condition, as shown in **Figure 6c**. The more advanced onset of LTHR than that of DME results from the higher T_o. As **Figure 6a** shows, the LTHR accelerates the temperature rise towards the end of the compression stoke. Therefore, T_o has to be reduced to achieve the same CA50.

5. Summary

HCCI is an alternative engine combustion process with potential for efficiencies as high as compression ignition (CI) engines while producing ultra-low particulate matter (PM) and nitrogen oxides (NO_x) emissions. HCCI engines operate on the principle of having a dilute premixed charge as like SI engines, which reacts and combusts throughout the in-cylinder as it is compressed by the piston. As stated above, HCCI incorporates the best features of both SI and CI engines. As like in SI engines, the charge is well mixed, which minimizes particulate emissions, and as like in CI engines, the in-cylinder charge is compression ignited by piston without the throttling losses, which leads to high thermal efficiency. Experiments and analysis to date suggest that chemical kinetics dominates thermal autoignition in HCCI. Detailed chemical-kinetics approaches have the advantage of directly simulating all the chemical processes leading to autoignition in HCCI engine. Detailed chemical-kinetic mechanisms have been developed for a wide variety of fuels, including methane, dimethyl ether (DME), iso-octane, n-heptane and many others. These mechanisms capture reaction rate information for elementary reaction steps. In other words, they capture the collisions that convert on molecule to another. The advantage of detailed chemical kinetics is that the processes leading to ignition are directly modeled and processes such as low-temperature reactions (LTR), intermediate-temperature reactions (ITR) and high-temperature reactions (HTR) can be solved. Numerical calculations for HCCI are often conducted with lumped (single-zone model) chemical-kinetics models, which assume spatially uniform temperature, pressure and composition in a fixed-mass, variable volume reactor. For this chapter, a zero-dimensional single-zone engine model of 'CHEMKIN' in Chemkin-Pro is applied to investigating the autoignition and chemical-kinetic mechanisms of HCCI combustion for the fuels with various autoignition reactivity. This is done for four fuels: methane, dimethyl ether (DME), iso-octane and n-heptane. Methane and iso-octane are selected as the single-stage ignition fuel, and DME and n-heptane are selected as the two-stage ignition fuel. A detailed chemical-kinetic mechanism for methane and DME is Mech_56.54 (113 species and 710 reactions). For iso-octane and n-heptane, a detailed chemical-kinetic mechanism from Lawrence Livermore National Laboratory (1034 species and 4236 reactions) is used. The results show that methane and iso-octane only exhibit the main heat release, 'high-temperature heat release (HTHR)' by HTR. In contrast, both DME and n-heptane exhibit the first heat-release 'low-temperature heat release (LTHR)' associated with LTR before HTHR. Because the LTHR accelerates the temperature rise towards the end of the compression stroke, the initial temperature has to be reduced to achieve the same combustion phasing. For a given initial pressure, a lower initial temperature leads to higher charge density and thus the higher amount of fuel when φ_o is constant. Eventually, the higher amount of fuel is advantageous for increasing the power output of HCCI engines.

Abbreviations and nomenclature

BDC Bottom dead centre

CAI Controlled auto ignition

CI Compression ignition

http://dx.doi.org/10.5772/intechopen.70541

CO	Carbon monoxide
COV	Coefficient of variation
DME	Di-methyl ether
EGR	Exhaust gas recirculation
HC	Hydrocarbon
HCCI	Homogeneous charge compression ignition
HRR	Heat-release rate
HTHR	High-temperature heat release
HTR	High-temperature reaction
ITR	Intermediate-temperature reaction
LTHR	Low-temperature heat release
LTR	Low-temperature reaction
NO_x	Nitrogen oxides
NTC	Negative temperature coefficient
PM	Particulate matter
RCM	Rapid compression machine
PRF	Primary reference fuel
SI	Spark ignition
TDC	Top dead centre
ZEV	Zero emissions vehicle
CA50	50% Burn point
To	Initial temperature
Po	Initial pressure
φ	Fuel/air equivalence ratio
φo	Initial fuel/air equivalence ratio
α	Characterisitic decomposition time
τ	Ignition delay time

Author details

Dongwon Jung

Address all correspondence to: great0526@gmail.com

Graduate School of Science and Technology, Keio University, Japan

References

[1] Reitz RD. Directions in internal combustion engine research. Combustion and Flame. 2015;**160**(1):1-8. DOI: 10.1016/j.combustflame.2012.11.002

[2] National Research Council. Assessment of Fuel Economy Technologies for Light-Duty Vehicles. Washington, DC: The National Academies Press; 2011. p. 232. DOI: 10.17226/12924

[3] Environmental Protection Agency. Regulations for Emissions from Vehicles and Engines [Internet]. [Updated: 30 March 2017]. Available from: https://www.epa.gov/regulations-emissions-vehicles-and-engines [Accessed: 20 June 2017]

[4] California Environmental Protection Agency Air Resources Board (CARB). Low-Emission Vehicle Regulations and Test Procedure [Internet]. [Updated: 22 August 2016]. Available from: https://www.arb.ca.gov/msprog/levprog/test_proc.htm [Accessed: 20 June 2017]

[5] Stocker TF, Qin D, Plattner G-K, Tignor M, Allen SK, Boschung J, Nauels A, Xia Y, Bex V, Midgley PM. Climate Change 2013: The Physical Science Basis. Contribution of Working Group I to the Fifth Assessment Report of the Intergovernmental Panel on Climate Change. United Kingdom and New York, NY. USA: Cambridge University Press; 2013. p. 1535. DOI: 10.1017/CBO9781107415324

[6] California Environmental Protection Agency Air Resources Board. Zero Emission Vehicle (ZEV) Program [Internet]. [Updated: 18 January 2017]. Available from: https://www.arb.ca.gov/msprog/zevprog/zevprog.htm [Accessed: 20 June 2017]

[7] Reif K, editor. Gasoline Engine Management. Mumbai, India: Springer Vieweg; 2015. p. 336. DOI: 10.1007/978-3-658-03964-6

[8] Reif K, editor. Diesel Engine Management. Mumbai, India: Springer Vieweg; 2014. p. 370. DOI: 10.1007/978-3-658-03981-3

[9] Zhao H. HCCI and CAI Engines for the Automotive Industry, Cambridge. England: Woodhead Publishing; 2007. p. 544. DOI: 10.1533/9781845693541

[10] Saxena S, Bedoya I. Fundamental phenomena affecting low temperature combustion and HCCI engines, high load limits and strategies for extending these limits. Progress in Energy and Combustion Science. 2013;**39**(5):457-488. DOI: 10.1016/j.pecs.2013.05.002

[11] Zhao F, Asmus TW, Assanis DN, Dec JE, Eng JA, Najt PM. Homogeneous Charge Compression Ignition (HCCI) Engines: Key Research and Development Issues. Pennsylvania. USA: Society of Automotive Engineers; 2003. p. 658

[12] Kee RJ, Rupley FM, Miller JA. CHEMKIN-III: A Fortran Chemical Kinetics Package for the Analysis of Gas Phase Chemical and Plasma Kinetics. Sandia National Laboratories Report No. SAND96-8216. 1996

[13] ANSYS CHEMKIN-PRO 18.0, ANSYS Reaction Design: San Diego, 2017.

[14] Heywood JB. Internal Combustion Engine Fundamentals. New York, USA: McGraw-Hill; 1988. p. 930

[15] Japan DME Association. DME Handbook. Tokyo, Japan: Ohmsha; 2007. p. 605

[16] International Programme on Chemical Safety. IPCS INCHEM [Internet]. Available from: http://www.inchem.org/ [Accessed: 20 June 2017]

[17] Burke U, Somers K, O'Toole P, Zinner C, Marquet N, Bourque G, et al. An ignition delay and kinetic modeling study of methane, dimethyl ether, and their mixtures at high pressures. Combustion and Flame. 2015;**162**(2):315-330. DOI: 10.1016/j.combustflame.2014.08.014

[18] Lawrence Livermore National Laboratory. Primary Reference Fuels (PRF): Iso-Octane/N-Heptane Mixtures [Internet]. Available from: https://combustion.llnl.gov/archived-mechanisms/surrogates/prf-isooctane-n-heptane-mixture [Accessed: 20 June 2017]

[19] Curran HJ, Gaffuria P, Pitza WJ, Westbrook CK. A comprehensive modeling study of iso-octane oxidation. Combustion and Flame. 2002;**129**(3):253-280. DOI: 10.1016/S0010-2180(01)00373-X

[20] Curran HJ, Gaffuri P, Pitz WJ, Westbrook CK. A comprehensive modeling study of n-heptane oxidation. Combustion and Flame. 1998;**114**(1-2):149-177. DOI: 10.1016/S0010-2180(97)00282-4

[21] Mansfield AB, Wooldridge MS, Di H, Hed X. Low-temperature ignition behavior of iso-octane. Fuel 2015;**139**(1):79-86. DOI: 10.1016/j.fuel.2014.08.019

[22] Warnatz J, Dibble RW, Maas U. Combustion: Physical and Chemical Fundamentals, Modeling and Simulation, Experiments, Pollutant Formation. 4th ed. New Jersey, USA: Springer Verlag; 2006. p. 378

[23] Chevalier C, Warnatz J, Melenk H. Automatic generation of reaction mechanisms for the description of the oxidation of higher hydrocarbons. Berichte der Bunsengesellschaft für Physikalische Chemie. 1990;**94**(1):1362-1367. DOI: 10.1002/bbpc.199000033

[24] Ando H, Sakai Y, Kuwahara K. Universal Rule of Hydrocarbon Oxidation. In: SAE Technical Paper 9 January 200948; April 20-23. Michigan, USA. Pennsylvania, USA: Society of Automotive Engineerings; 2009. DOI: 10.4271/2009-01-0948

[25] Smith JR, Green RM, Westbrook CK, Pitz WJ. An experimental and modeling study of engine knock. Proceedings of the Combustion Institute. 1985;**20**(1):91-100. DOI: 10.1016/S0082-0784(85)80492-6

Chapter 4

New Materials to Solve Energy Issues through Photochemical and Photophysical Processes: The Kinetics Involved

Tatiana Duque Martins,
Antonio Carlos Chaves Ribeiro,
Geovany Albino de Souza,
Diericon de Sousa Cordeiro, Ramon Miranda Silva,
Flavio Colmati, Roberto Batista de Lima,
Lucas Fernandes Aguiar, Leandro Lima Carvalho,
Renan Gustavo Coelho S. dos Reis and
Wemerson Daniel C. dos Santos

Additional information is available at the end of the chapter

http://dx.doi.org/10.5772/intechopen.70467

Abstract

Kinetic rates of energy production are extremely controlled by the competing processes that occur in systems capable of energy transfer. Besides organic and inorganic compounds already known as electronically actives, supramolecular systems can be thought to form energy transfer complexes to efficiently convert, for instance, light into electricity and the mechanisms for that can be of any kind. Photophysical and photochemical processes can simultaneously occur in such systems to provide energy conversion, by competing mechanisms or collaborative ones. Thus, to investigate the kinetic rates of each process and to understand the dynamics of the electronic excited states population and depopulation in strategically structured materials, can offer important tools to efficiently make use of this not always so evident power of supramolecular materials. In this chapter, we present the state-of-the-art of the use of photophysical processes and photochemical changes, presented by new materials and devices to provide a control of energy transfer processes and enable distinct applications, since energy conversion to sensing and imaging techniques to material characterization.

Keywords: photochemistry, photophysics, kinetics, excited states lifetime, electronic energy transfer, nanomaterials, energy conversion

1. Introduction

In nature, there are a number of processes indispensable for life maintenance that begins with light absorption. From this starting point, several chemical changes, ranked by probabilities of occurrences, are triggered to give a product. In this process, molecular photophysical and photochemical processes occur simultaneously, competing to each other for the excess energy. On the other hand, these competing processes are also collaborating to each other, since they occur through electronic excited state reactants that originate electronic excited state intermediates. Based on the structures and characteristics of these excited electronic states intermediates, new mechanisms can be proposed, yet involving dissociations, isomerization, bond cleavages, nevertheless, taking into account that these excited species present peculiar electronic distribution and, therefore, involve photophysical activation and deactivation mechanisms, that arise from their interaction with light, all governed by new and challenging kinetic laws. In this sense, the peculiar characteristic of the kinetic laws involved in molecular photophysical processes is that electronic excited species that can be reached by light absorption are considered unstable, and to achieve a more stable electronic configuration, excess energy is liberated by radiative and/or non-radiative unimolecular decays.

The photophysical processes that occur immediately following the light absorption aim to ensure the mechanisms to achieve the best energetic configuration to: (1) lead to the reactive excited intermediate, from which the photochemistry can occur or (2) achieve the faster way to release the excess energy and to retrieve the initial reactant. They can all be defined in a Jablonski diagram [1] (**Figure 1**) and their corresponding rate expressions can be obtained from there.

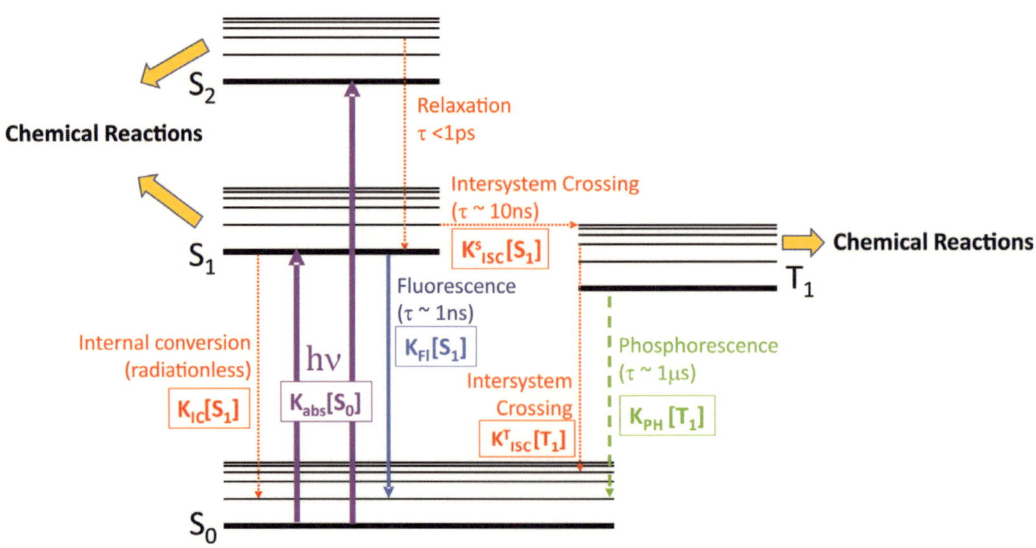

Figure 1. Jablonski diagram presenting the major radiative and non-radiative processes and their rates.

http://dx.doi.org/10.5772/intechopen.70467

2. Photophysical processes

2.1. Absorption

The initial photophysical process that gives rise to excited states from where every photophysical and subsequent photochemical processes occur in the radiative absorption of photons to promote an electron to a higher electronic energy state. The accessed excited state is determined by selection rules that involve symmetry and spin conservation, existence of a dipole moment and must occur to an ideal vibrational mode wavefunction in the excited state overlapped in some extent with the low energy vibrational mode of the ground electronic state, enabling some probability of transition, as predicted by the Franck-Condon principle. The magnitude of this overlap influences the moment transition in absorption and every other photophysical processes [2]. The expression that describes the transition is:

$$M = \int \psi *_{el,higher} \mu \psi_{el,lower} d\tau \int \psi *_{vib,higher} \psi_{vib,lower} d\tau \tag{1}$$

Where the second integral is the overlap integral. From this expression, it is evident that there must be a probability of a wavefunction from a lower electronic state to absorb enough energy to be converted in another wavefunction that describes a higher electronic state and that if there is no overlap between the vibrational states expected to be involved in the transition, then the electronic transition is forbidden. It evidences the *vibronic* nature of the electronic state, in which electronic states are coupled to vibrational states. **Figure 2** presents the Franck-Condon absorption from the ground electronic state to a vibronic state of higher energy.

The absorption process populates electronic excited states from where all deactivation processes will occur. The most significant photophysical deactivation processes are:

2.2. Fluorescence

The photophysical process in which the electronic excited state is radiatively deactivated, involving singlet excited and ground states, is the fluorescence. It spontaneously occurs from the singlet excited state of lower energy, as predicted by Lewis and Kasha [3], through the emission of a photon and the energy involved in this process is similar to the absorbed energy, if no other competing process of deactivation occurs. It occurs very rapidly in a timescale that depends on the system identity but between 10^{-6} and 10^{-10} seconds for several organic compounds. If longer timescales are observed, it may evidence the occurrence of another process that results in a similar spectrum, but occurs after some other photophysical deactivation processes that populate the singlet electronic state of lower energy. This is the *delayed fluorescence* and it only can be distinguished from the fluorescence by time-resolved measurements.

2.3. Phosphorescence

Phosphorescence is a radiative deactivation process characterized by a red-shift of the emission spectrum. It is a process that occur from an electronic excited state with less energy than that from where fluorescence occurs. In fact, it occurs from a triplet electronic state with less energy

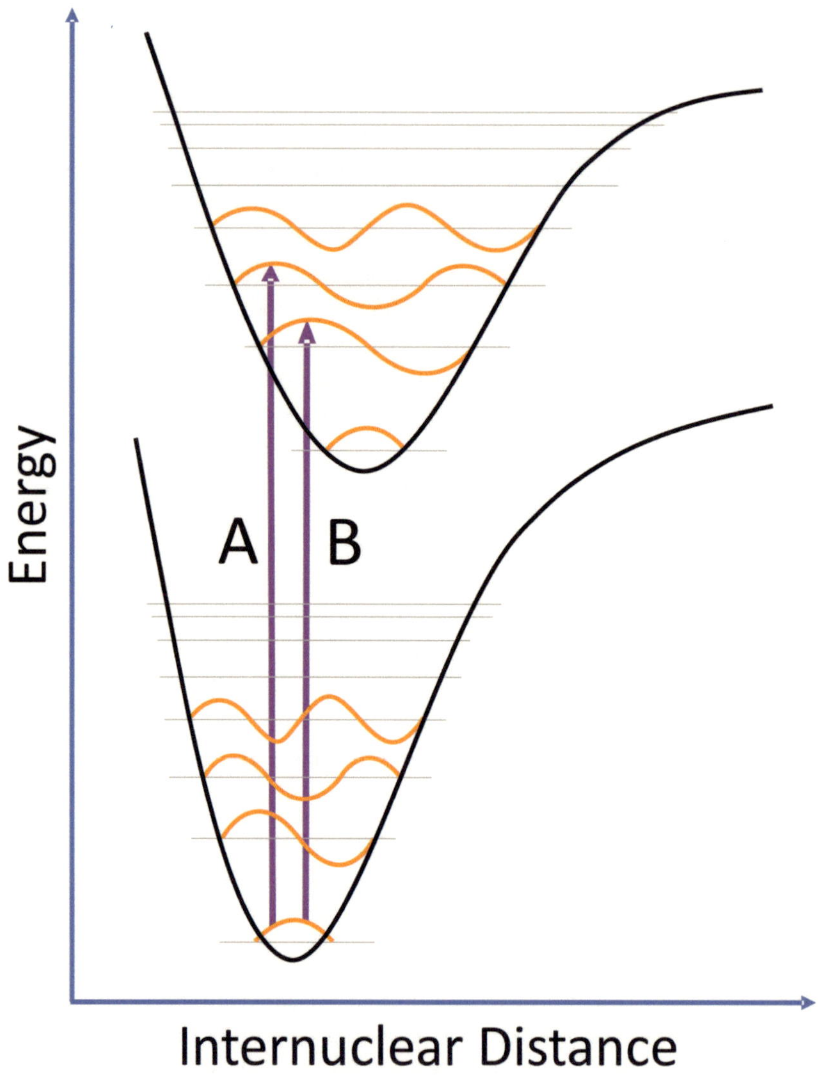

Figure 2. Franck-Condon vibronic absorption from the electronic ground state to an excited state: From the lowest vibrational state (v_0) in the ground state to A) the lowest vibrational state (v_0) in the excited electronic state and B) to a higher vibrational state (V_4) in the excited electronic state.

than the singlet electronic excited state of lower energy. Since spin changes are forbidden in electronic transitions, this is a process that occurs only if relaxation in the spin selection rule occurs, provided by spin-orbit coupling derived from the coupling of the electron spin motion with its orbital motion. Due to that prohibition, this is a very slow process, taking from 10^{-6} seconds to minutes or even hours to occur [2].

2.4. Vibrational relaxation

The process of releasing the energy given by the absorption of a photon as kinetic energy is the vibrational relaxation. It involves the conversion of a vibrational mode within an electronic state to another vibrational mode within the same electronic state. This process is very fast, taking around 10^{-14}–10^{-11} seconds. It usually takes place immediately following absorbance and, since it occurs between vibrational levels, generally it does not result in electronic level changes [1, 2].

2.5. Internal conversion

A non-radiative process that promotes the conversion of a singlet electronic excited state of higher energy into another singlet electronic state of lower energy is the internal conversion. It can involve any two singlet states and, when occurring between the singlet electronic excited state of lower energy and the singlet ground state, it competes with fluorescence, being one reason for a decrease in fluorescence quantum yield. It occurs rapidly with release of kinetic energy [1, 2].

2.6. Intersystem crossing

The non-radiative process of conversion of an electronic excited singlet state into a triplet one through an isoenergetic process is the intersystem crossing. This is a very slow process, because it is forbidden by spin multiplicity selection rules and it only takes places if an effective spin-orbit coupling occurs [1, 2].

These radiative and non-radiative processes are unimolecular, involving only the electronic states of a single molecule. Nevertheless, there are several other bimolecular processes, characterizing energy transfer processes or even chemical reactions.

3. Energy transfer processes

Energy transfer can occur between similar molecules or distinct compounds and the way they interact will define the more appropriate transfer mechanism for each case. Depending on the mechanism and the energetic characteristics of the energy transfer red, the transfer can be classified as [1]:

1. Hole transfer: When a positively charged molecule interacts with another molecule to achieve its energetic equilibrium and resulting in the second molecule to present the positive charge.

$$D^{+} + A \rightarrow D + A^{+} \quad \textbf{hole transfer} \tag{2}$$

2. Electron transfer: Similarly, if a negatively charged molecule interacts in some way with another neutral molecule to result in the second molecule now as negatively charged.

$$D^{-} + A \rightarrow D + A^{-} \quad \textbf{electron transfer} \tag{3}$$

3. Energy transfer: When the interaction between molecules, one of them in the electronic excited state and the other occupying the electronic ground state results in the second

molecule occupying the excited state and the initially excited molecule in the electronic ground state.

$$D^* + A \rightarrow D + A^* \quad \textbf{energy transfer} \tag{4}$$

The energy transfer mechanisms involve an entity which presents the excess energy, defined as donor (D) and an entity that can receive this excess energy, defined as acceptor (A). They are classified as radiative or a non-radiative process, depending on the occurrence of the luminescent emission from the donor.

3.1. Radiative energy transfer

The donor in the electronic excited state relaxes to radiatively release its excess energy. Thus, fluorescence (or phosphorescence) needs to occur to promote the energy transfer through the absorption of the fluorescence of the donor by the acceptor [4]. It is known as the trivial energy transfer mechanism and it is enabled by the overlap between the absorption spectrum of the acceptor with the luminescence spectrum of the donor. It does not require that donor and acceptor be in the same environment and it is independent of the luminescence lifetime of the donor and depends on the concentration of the acceptor ([A]), the quantum yield of the donor (φ_{eD}) and the molar extinction coefficient of the acceptor (ε_A).

$$Rate = k_{fl}^{D} P_{abs}^{A} [D^*] \quad \text{with} \quad P_{abs}^{A} \propto [A] l \int_0^\infty F_D(\nu) \varepsilon_A(\nu) d\nu \tag{5}$$

Scheme in **Figure 3** presents the trivial mechanism of energy transfer.

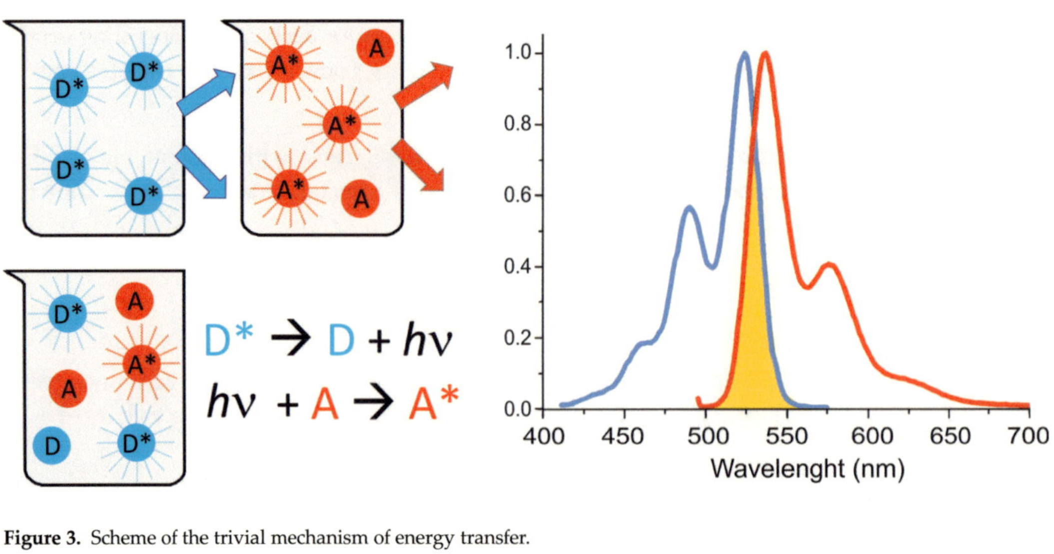

Figure 3. Scheme of the trivial mechanism of energy transfer.

http://dx.doi.org/10.5772/intechopen.70467

3.2. Non-radiative energy transfer

Inversely to the trivial mechanism, non-radiative energy transfer mechanisms are strictly dependent on the luminescence lifetime of the donor, since it only occurs while the donor is in its electronic excited state. It needs the formation of a collision complex between the donor and the acceptor and the energy transfer occurs with the right molecular distance:

$$\mathbf{A + D^* \rightarrow [AD^*] \rightarrow [A^*D] \rightarrow A^* + D} \quad (6)$$

Its rate is given by the magnitude of the transition moment between the electronic wavefunction that describes the collision complex before and after the transfer from the donor to the acceptor:

$$k = \frac{2\pi}{\hbar}\left|\int \Psi^{*}_{A^*D} H' \Psi_{AD^*}\right|^2 \rho(E) \quad (7)$$

Where $\Psi_{AD^*} = \Psi_A \Psi_{D^*}$ is the complex wavefunction before the energy transfer and $\Psi_{A^*D} = \Psi_{A^*}\Psi_D$ is the wavefunction that describes the complex after the energy transfer.

Depending on the nature of the energy transfer, the intermolecular distance and the similarity of excited state energies, they can occur by a resonant mechanism called Forster resonance energy transfer (FRET) or based on the electron exchange called Dexter energy transfer.

3.3. Forster resonant energy transfer

Energy transfer that occur in a rate similar to the donor fluorescence lifetime initially involves a Coulombic interaction between the electronic excited state of the donor and the electronic ground state of the acceptor that evolves to interaction of the acceptor electronic excited state with the donor ground state. These Coulombic interactions are only possible due to the energy proximity of the emission of the donor and the absorption of the acceptor, enabling a virtual energy transfer, in which absorption and emission of the energy occur simultaneously. Because the Coulombic interactions between the electronic states of both donor and acceptor occur during the donor fluorescence lifetime and they are predominant and represent the influence of the dipole-dipole interaction, they are dependent on the inter-species distance by a factor of r^{-3}. The probability of occurrence of the energy transfer, then, is proportional to the square of the distance, hence r^{-6}. The rate of the energy transfer is given by the Forster expression [4]:

$$k_{DA} = \frac{9000k^2 \ln 10}{128\pi^5 n^4 N_A \tau_{DA} r^6} \int \frac{F_D(\tilde{\nu})\varepsilon_A(\tilde{\nu})}{\tilde{\nu}^4} d\tilde{\nu} \quad (8)$$

Where k^2 is the relative orientation of the dipole of the donor and the acceptor, F_D is the intensity of fluorescence of the donor, ε_A is the acceptor coefficient of extinction, τ_{DA} is the donor fluorescence lifetime in the presence of the acceptor and r is the distance between donor and acceptor.

When the probability of occurrence of non-radiative energy transfer is 50%, a critical distance, called Forster radius, is reached and it is defined as the distance in which the transfer rate, k_{DA}, is equivalent to the donor fluorescence lifetime, when in the absence of the acceptor, $\tau_D{}^{-1}$:

$$k_{DA} = \frac{1}{\tau_D}\left(\frac{R_0}{r}\right)^6 \tag{9}$$

The critical distance is much longer than the bond distances and the energy transfer is said to be a long-distance energy transfer.

3.4. Dexter energy transfer

The mechanism of electronic energy transfer that involves the electron transfer between the electronic excited state of the donor to the unoccupied excited state of the acceptor, simultaneously to the transfer of an electron of the electronic ground state of the acceptor to the poorly occupied electronic ground state of the donor, characterizing an electron exchange mechanism is the Dexter energy transfer. Since it is an exchange interaction, it needs an overlap between the wavefunctions of the donor and the acceptor to occur.

The rate of the electron exchange is proportional to the ratio between the donor-acceptor distance and the sum of their Van der Waals radii.

$$k_{T\,(exchange)} \propto \exp(\,2r_{DA}\,/\,L) \tag{10}$$

The donor-acceptor distance, in this case, is short, corresponding to distances of a complex formation. These mechanisms are illustrated in **Figure 4**.

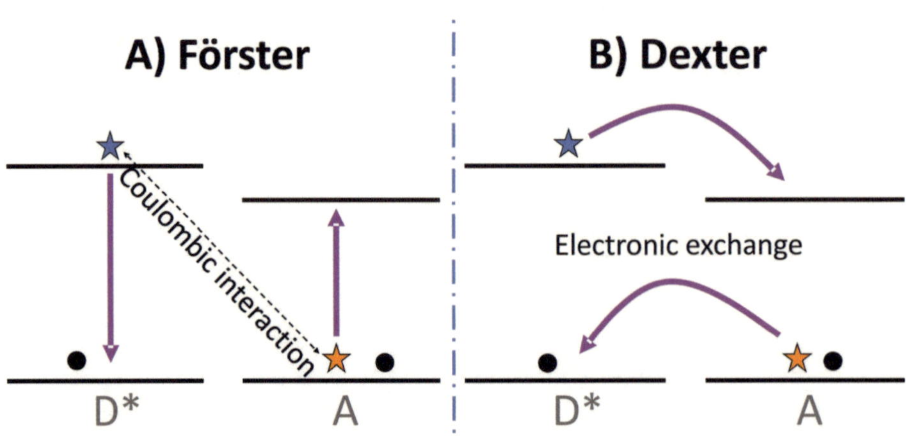

Figure 4. Diagrams illustrating the (A) Forster resonant energy transfer and (B) Dexter energy transfer mechanisms.

http://dx.doi.org/10.5772/intechopen.70467

4. Energy transfer complexes

Non-radiative energy transfer mechanisms involve the formation of energy transfer complexes. In most cases, these complexes are formed by collision; thus, their kinetics of formation is governed by diffusion rates and is dependent on the molecule-environment interactions. Its

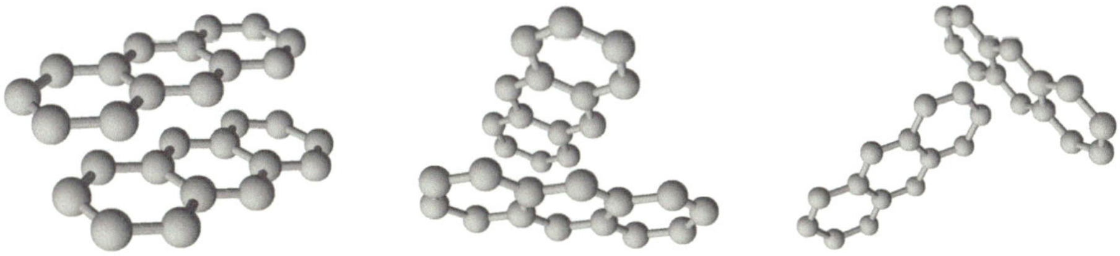

Figure 5. Excimer configurations.

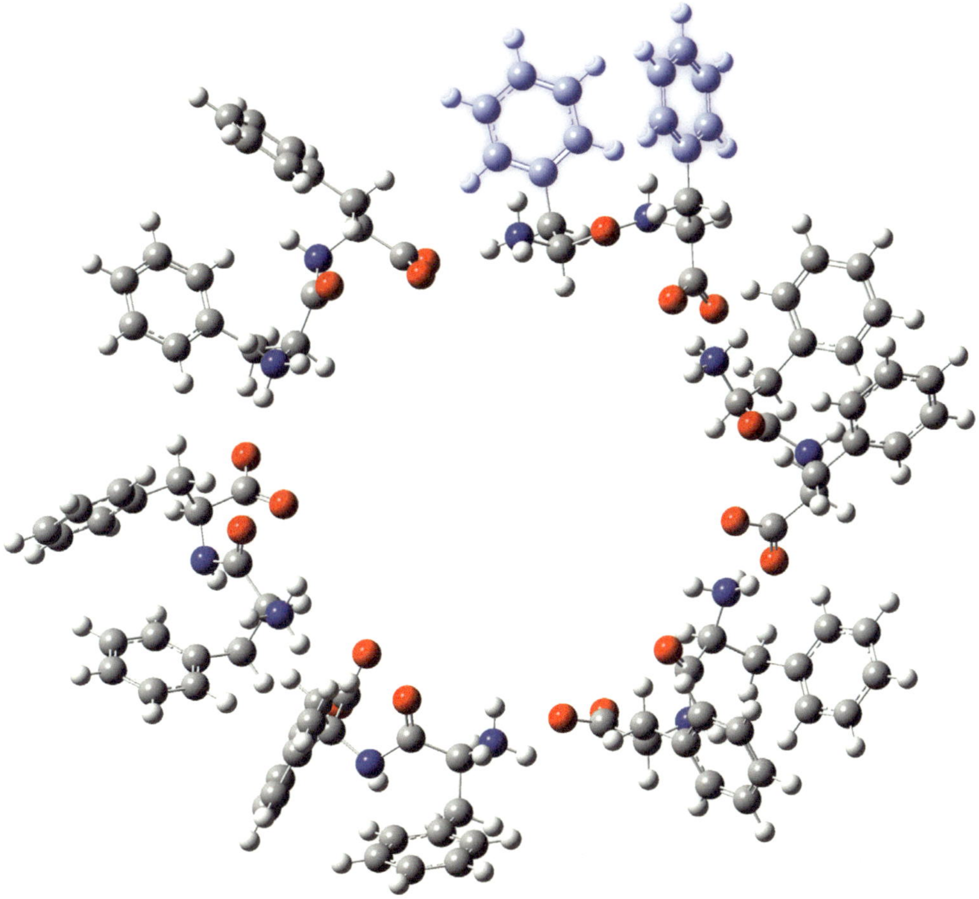

Figure 6. Supramolecular diphenylalanine hexagonal crown forming an energy transfer complex upon absorption of the phenyl groups of a single peptide.

mandatory exigence is to have one of the molecules involved in the complex formation in the electronic excited state. The success of collisions will give the number of intermediates in the excited states that present the ideal characteristics for energy transfer. These excited state complexes are classified depending on the identity of their components [2–4]:

1. *Excimers* are the excited state complexes that are formed by two similar compounds. They present the same absorption electronic spectra as the isolated molecules, but emission spectra broader and red-shifted than the emission expected for the isolated molecule. The emission spectrum is the result of the emission of a new compound, the complex, formed during the excited state of the molecule that absorbed the electromagnetic radiation and is formed by collision. Excimers present several distinct orientations, from the totally overlapped orientation, called sandwich excimer, to some partially overlapped and the t-shaped excimer. **Figure 5** presents these configurations.

2. *Exciplexes* are the complexes formed by distinct compounds, with one of them being at the electronic excited state. They are also governed by diffusion rates, but in a very specific manner, since it depends on efficient simultaneous collisions. Their absorption spectra are similar to that observed for the isolated absorber, but the emissions are very difficult to predict, since several competing pathways of deactivation, with kinetics influenced by the environment and the interaction forces acting to keep the exciplex together, during the excited state of the complex. This is the case of exciplexes involved in supramolecular photochemical reactions, as exemplifies in **Figure 6**.

5. From photophysical to photochemical processes

All these photophysical processes modulate the energy and the characteristics of the intermediates prior to the occurrence of photochemical modifications. They occur in typical amounts of time; thus, light absorption is the determining step and it takes femtoseconds (10^{-15} seconds) to occur. The radiative deactivation of the lowest excited state to reach the ground state is the fluorescence, which occurs in nanoseconds (10^{-9} seconds) timescale; its occurrence informs about the electronic excited state lifetime and, therefore, about its stability. If it is long enough, several processes can occur and the radiative deactivation is not observed or its yield is diminished. From there, reactive intermediates can be formed in the excited state and, if funnels or interconversion situations are avoided by, for instance, guaranteeing that the energy barrier is too high to be superposed, then the final product, result of all photophysical and photochemical processes that occur during the lifetime of the electronic excited state, is the excited product. The ground state product is obtained when the excess energy is released as radiative emission [3].

Nevertheless, if the energy barrier is superposed and funnels are formed, the reactive excited state intermediate cannot be formed and the chemical reaction occurs in the ground state. These events can be summarized in **Figure 7**.

The rate constants and the probabilities of these processes determine which path can lead to the product formation. To describe the excited states and the changes that occur to yield the

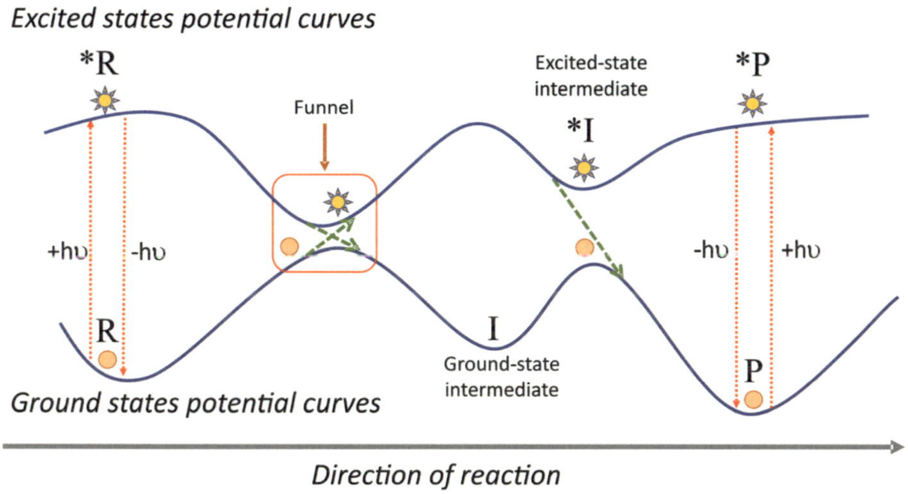

Figure 7. Potential curves of ground and electronic excited states of a photochemical reaction. (R) is the ground state reactant, (*R) is the excited reactant, (I) is the ground state intermediate of a reaction, (*I) is the excited state intermediate, (*P) is the excited state product and (P) is the final product of the overall process.

product is the key to perform any kind of reaction control and to choose all the experimental conditions that satisfy the reaction requirements. The rate constants, the intermediate formation and structures, the reasons for interconversions, energy migrations and excited states deactivation are crucial to exert any sort of reaction control. For that, the kinetic laws of excited state intermediate formation, the characteristics of funnels and the difference between thermal and photo-activated chemical reactions and the kinetics involved in energy transfer processes must be scrutinized. As showed by Soboleva et al. [5], to describe the electronic excited states lifetimes is very important to even propose mechanisms for charge transfer in supramolecular systems. In their work, they showed that electron transfer kinetics can be monitored by time-resolved luminescence quenching measurements of a chromophore in the presence of a quencher to describe the electron-transfer reactivity in sodium dodecyl sulfate (SDS) micellar systems. They observed that the mobility of the quencher is faster than the electron-transfer rate, which resulted in the conclusion that, in the cases where electron transfer between donor and acceptor is slower than the diffusion rate, the transfer is then controlled by reaction kinetics instead of by diffusion.

All these phenomena occur in a system of competition vs. cooperation, through intermediates and governed by probabilities of occurrence and rate constants, as they direct the mechanisms that are employed in a great number of applications. Examples are probing in imaging diagnosis, energy conversion and storage, data storage, photodynamic therapy, among several others.

6. State-of-the-art

Nowadays, photophysical and photochemical processes are perceptively and actively being applied in several areas of science and technology to promote a rapid and sustainable way to better everyone's life worldwide. Examples are the several uses of photochemistry kinetics in distinct processes and its application to new materials development, in special those for energy conversion and energy harvesting [6–11].

Recently, research into optoelectronic organic materials is being developed to describe new options with potential for applications in emissive devices, sensors and solar cells [7]. Although these materials have been successfully tested as part of these devices, they are numerous and a serious difficulty has been to determine which characteristics are determinant for a material to present a specific property and how to replicate that in others. The answer invariably has been found in determining the kinetics of deactivation of the electronic excited states and, therefore, of the photophysical properties and photochemical processes. The efficiency of a device containing organic electroluminescent compounds is strictly related to the efficiency of the exciton formation and, thus, it depends on the conjugation lengths [7], which determine the mechanisms of energy transfer among the material [12]. For instance, in their work, Arkan and Izadyar studied the mechanism of charge transfer and the rate of exciton formation and dissociation in dye-sensitized solar cells based on TiO_2/Si/porphyrins. They observed the rate of exciton formation/dissociation in metal-porphyrins, revealing the occurrence of an efficient charge transport in these systems.

Indeed, it is expected that efficient solar cells present great ability of exciton formation, efficient exciton transport and charge transport from the donor to the acceptor [13] to minimize the influence of the competitive processes such as exciton recombination that reduces the energy conversion efficiency [14].

Exciton formation is a driving force of the solar cell efficiency, which causes the exciton recombination to be an event that needs to be controlled. In several devices, recombination must be understood to be avoided to guarantee the highest efficiency. Many solar cells have been based on perovskite due to their ability of delivering efficiencies as high as 22% [15]. In their work, Dar et al. characterized the charge carrier recombination process that occurs in a bromide-based perovskite by measuring the transient absorption kinetics are several excitation intensities (5–100 μJ cm^{-2}). For that, they assumed that the carrier dynamics is mainly governed by bimolecular recombination, being expressed and decay kinetics:

$$dn/dt = \gamma(t)n^2 \tag{11}$$

Where, in disordered systems, the time-dependent recombination is approximately to [16]:

$$\gamma(t) = \gamma_0 t^{-\alpha} \tag{12}$$

That gives the carrier concentration kinetics: $1/n = -1/n_0 = \gamma_0\ t^{1-\alpha}/(1-\alpha)$, independent of the initial carrier density and, thus, independent of the excitation intensity.

Through this treatment, they identified the time-dependent recombination as a function of the morphology of the perovskite. They found that the polycrystalline perovskite structure presents grain boundaries that are physical obstacles for the carrier motion, which results in a decrease of the recombination rate. They were able to determine that the recombination rate constant is a consequence of the perovskite morphological inhomogeneity.

Recombination is an important mechanism of depopulation of the excited state, from which energy is generated. Controlling the exciton recombination has been a strategy for enhancing the solar cell efficiency, but it needs an accurate characterization of the kinetics of all competing processes of deactivation and, sometimes, it can lead to a poorly effective control of the recombination. Other strategies have been developed, focusing on enhancing the exciton formation, other than avoiding recombination. Many studies have demonstrated that processes such as multiple exciton generation in quantum dots and singlet exciton fission in molecular chromophores have greatly contributed to enhance the power conversion efficiency of devices such as solar cells and fuels cells. To carefully characterize, both processes had proven to consist of an embracing strategy to promote higher efficiencies. Beard et al. [17] studied the characteristics of the mechanisms multiple exciton generation [18] and singlet exciton fission [19, 20], searching for their similarities, in order to give enough information on how to improve the exciton formation in such devices, independently of the device configuration. They found that the two mechanisms are different, because in multiple exciton generation, two excitons are created in a single quantum dot whereas in singlet exciton fission, two species are electronically coupled to give rise to an overall singlet excited state that allows a transition from the singlet excited state to two coupled triplet excited states. In the former, there is spin conservation, in the latter, two triplets are created, each one presenting half the energy of the prime singlet excited state. Also different are their dynamics. Exciton multiplication, in both mechanisms, occurs very fast, nevertheless, the difference lies on lifetimes of the newly generated excitons. In exciton singlet fission mechanism, the new excited triplet states present lifetimes of microseconds, originated from singlet states with lifetimes of nanoseconds [19], whereas in multiple exciton generation, the excitons present lifetimes of picoseconds [21]. Despite these differences, they concluded that in solar cells, the enhancement in the efficiencies calculated considering both mechanisms are similar. They informed that there is still much work to be done regarding the solar cell structures to minimize non-radiative recombination and provide more efficiency to them, but solar cells with power conversion efficiency of over 30% can be easily obtained by multi-exciton generation. Also, Thompson et al. [22] showed that it is possible to achieve more efficient solar cells exploiting the singlet exciton fission mechanism, and Semonin et al. [23] achieved an increase in the external photocurrent efficiency of quantum dot solar cells exploiting the multiple exciton generation mechanism.

The photophysical processes that are responsible for the population of electronic excited states after the fast absorption of light by the absorber can be exploited for several imaginable applications. An example is the work of Wu et al. [24], where photolysis kinetics, quantum yield and bioavailability of a ketone (acetylacetone) during UV irradiation were investigated. They found that, after the absorption of UV light by the ketone, a series of photophysical processes overcame the photochemical reactions of decomposition. Interestingly, they observed that the energy transfer mechanisms that occur after the absorption of sunlight guarantee the high efficiency of the photochemical changes. Since the degradation products of the ketone after the photochemical reactions were similar to the metabolic products in biofermentation, they argue that the acetylacetone may be used in water treatment at the pre-treatment stage and may give some important information on the photochemical characteristics of several other β-diketones in water.

The energy transfer in organic systems can also be used to monitor distinct environments by enabling several mechanisms of tracking the changes in the electronic excited states involved in the photophysical or photochemical processes. Sensing and imaging are, therefore, ways to collect information on distinct environments.

In our research group [25], we have focused on the proposal of new materials that are able to efficiently form energy transfer complexes and give rise to new photophysical characteristics that are very sensitive to specific environmental changes. An example is a new material based on supramolecular structures of a dipeptide, diphenylalanine, composing an exciplex with a chromophore, coumarin. In distinct proportions, this system was able to modulate the coumarin sensibility to $O_{2(g)}$ dissolved in water, presenting distinct fluorescence spectra from that expected for coumarin, which was a result of the energy transfer complex formation and the new electronic excited states that resulted from the interactions between the components. Wang et al. [8], on the other hand, developed a method for monitoring photochemical reaction kinetics, presenting spatial resolution, the laser-excited muon pump-probe spin spectroscopy (photo-μSR). With this, they expected to monitor the dynamic of excitations and to explore the mechanism of photophysical and photochemical processes. Using pentacene as subject, they temporally and spatially mapped these processes at the single-carbon level and observed that the photochemical reactivity of a specific carbon atom is modified in the presence of a specific excited state.

Energy conversion can also be based on hole transfers or proton transfers and can involve photophysical processes, photochemical reactions or both processes in a collaborative way. Elbin and Bazan [7] proposed a new electron-deficient compound based on three-coordinate boryl substituents adjacent to highly conjugated distyrylbenzene derivative (DSB) or poly (aryleneethynylene)s (PAE). In these materials, boron atom provides a vacant p_z-orbital that confers them a strong electron acceptor character, enabling a significant delocalization. They showed that due to the distinct photophysical characteristics of the constituents, the excited state migration by FRET is modulated and, depending on the substituent, light of distinct colors are emitted from these systems. Based on that, they believed that these materials can find application in displays.

Also based on hole transfer to promote energy conversion is the electrochemical energy conversion in a system called fuel cell. It consists of an additional way for chemical energy conversion, without photocatalytic effect. It is an electrochemical system which converts chemical energy into electricity through the oxidation of a fuel [26, 27], which takes place in the anode of the cell, and the reduction of the oxygen from atmosphere in the cathode. Some of these fuel cells are classified by temperature operation [28], especially, Proton Exchange Membrane Fuel Cells (PEMFCs) work at low temperatures (from room to 100°C) [29] with a Nafion® membrane electrolyte. Low temperatures requires a very active catalyst in the electrodes, usually being platinum (Pt) [30]. A direct ethanol fuel cell (DEFC) is a very attractive electrochemical energy converter [31], and its unitary fuel cell scheme is shown in **Figure 8**. The fuel is supplied into the anode side and the air (or pure O_2(g)) is supplied into the cathode. The electrolyte carries protons from the anode to the cathode and the electrons are availed at an external electrical circuit to produce work.

http://dx.doi.org/10.5772/intechopen.70467

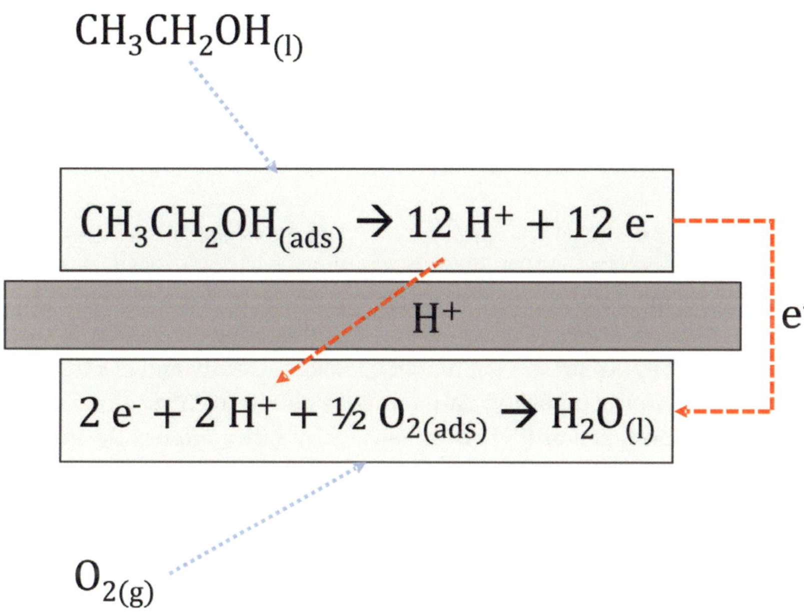

Figure 8. Scheme of direct ethanol fuel cell.

Ethanol is inserted into the fuel cell, adsorbs at electrode surface and is oxidized as shown in **Figure 8**.

$$CH_3CH_2OH \rightarrow CO_2 + H_2O \tag{13}$$

While oxygen from air is reduced:

$$O_2(g) + H^+(aq) + e^- \rightarrow H_2O(l) \tag{14}$$

Which gives the overall reaction of the direct ethanol fuel cell (admitting complete ethanol oxidation reaction):

$$CH_3CH_2OH(l) + O_2(g) \rightarrow H_2O(l) + CO_2(g) + \text{energy} \tag{15}$$

With the energy being mostly electrical work and heat. The electric work is dependent on the potential difference between cathode and anode: the larger the difference, the bigger the electrical work. Redox kinetics, thus, influences this amount of energy conversions, by inducing the number of electrons that are injected into the electrical circuit, resulting in electrical current.

At the anode, the ethanol adsorbs on electrode and the oxidation is characterized by the dehydrogenation. Some studies with Fourier transformed infra-red (FTIR) in situ [32], differential electrochemical mass spectroscopy (DEMS) [33, 34] show that the main products from electrochemical ethanol oxidation reaction, on Pt-based catalysts, are acetic acid and formaldehyde [35]. The electric work produced by direct ethanol fuel cell depends on the number of electrons that circulate at electrical circuit and the number of electrons generated by the redox reaction. Thus, the kinetic of ethanol oxidation reaction limits fuel cell performance.

Rightmire et al. [36] studied the ethanol oxidation reaction on Pt in acidic media and showed the determining step of the reaction is formaldehyde formation. Moreover, Hitmi et al. [37] showed that the rate of formation of acetaldehyde is larger than acetic acid formation from ethanol oxidation reaction. The formation of acetic acid from acetaldehyde depends on the adsorption of acetaldehyde on electrode surface, as proposed by Podlovchenko et al. [38].

$$CH_3CH_2OH_{sol} \rightharpoonup CH_3CH_2OH_{ads} \rightharpoonup CH_3CHO_{ads} \rightleftharpoons CH_3CHO_{sol} \quad (16)$$

The main problem of the catalysts is the poisoning effect by carbonaceous products from ethanol oxidation reaction strongly adsorbed on Pt. Nowadays, research is focused on the development of new catalytics presenting higher chemical stability and electrochemical kinetic rates. There are several works reporting Sn-modified Pt electrocatalyst as a more active material for ethanol oxidation reaction [39]. There are many other interesting materials, such as PtRh [40], PtMo [41] and PtPd [42], but better performances of DEFC were observed employing PtSn at the anode, which effects the kinetic of ethanol oxidation [43, 44].

Figure 9 shows the linear sweep voltammetry obtained for the ethanol oxidation on Pt electrocatalysts. FTIR were collected *in situ* with electrode polarization in ethanol solution. Pt

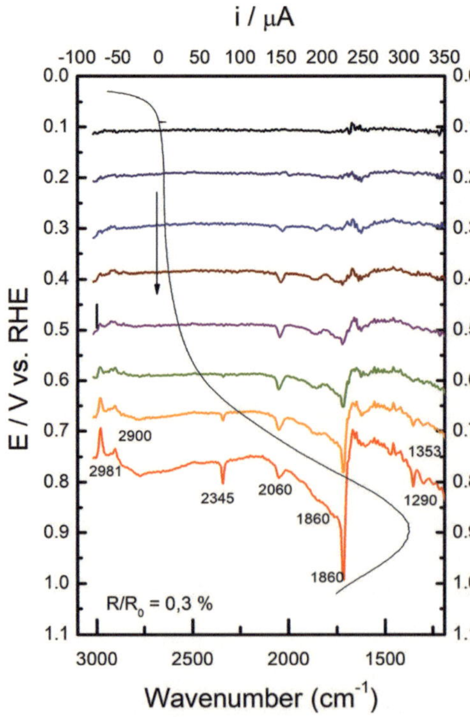

Figure 9. Linear sweep voltammetry and FTIR registered on Pt catalysts in 0.5 mol L^{-1} H_2SO_4 and 0.5 mol L^{-1} CH_3CH_2OH, at room temperature, v(lsv) = 1 mV^{-1} and FTIR measurements carried out in a mixture of 0.1 $HClO_4$ (mol L^{-1}) and CH_3CH_2OH (0.1 mol L^{-1}).

was polarized at 0.05 V vs. Reversible hydrogen electrode (RHE) and potential scan was set to 1.0 V at 1 mV s^{-1}, and the current in μA at top axis. The FTIR were collected at distinct electrode polarizations on steps of 0.1 V. The negative bands correspond to the formation of chemical species and positive bands correspond to consumption of adsorbed chemical species. The band at 2345 cm^{-1} refers to CO_2 formation [45] and it is observed only above 0.6 V vs. Reversible Hydrogen Electrode (RHE). The peak at 1860 cm^{-1} corresponds to COOH deflection [45] observed at 0.2 V, which suggests the fast formation of acetic acid on Pt, in acid solution and a difficulty to generate CO_2, which indicates complete ethanol oxidation. Peaks at 2981 and 2900 cm^{-1} correspond to CH_2 and CH_3 stretching, resulting from ethanol consumption. The peaks at 1715, 1353 and 1290 cm^{-1} correspond to the formation of aldehydes and carboxylic acids, such as acetaldehyde and acetic acid [32, 37].

Thus, the conversion of chemical energy into electrical energy depends on the potential and the kinetics of the reactions; the development of new materials for a better exploitation of fuel is, then, limited by the characteristics of the electrochemical reactions kinetics.

7. Conclusion

To understand the kinetic rates and laws of the dynamic processes of the energy transfers that involve the interaction between compounds, through the electronic excited states and the characteristics of the excited states is crucial to determine the applications, specially in energy conversion. Also, photochemical processes can be greatly exploited to cause the modifications in the materials that enable their ability of energy transfer. Regarding to this, the rate constants of the photochemical reactions determine the paths that yield products and they are strictly related to the electronic excited states involved in the photochemical processes. If rate constants, intermediate structures and their mechanisms of formation and the energetic balance involved in each change, it is possible to achieve the desired reaction control through experimental conditions control. New materials, capable of distinct electronic processes that can influence photophysical and photochemical processes, are of great interest, nowadays. They become more and more specific and selective, aiming higher efficiencies of energy conversion, as well as faster and sustainable ways to promote degradation of pollutants. Also, as energy conversion in fuel cells, depends on the kinetic rates of electron generation, the development of material for complete oxidation reaction of ethanol would disseminate its usage. This means that there are no limits to develop new materials with properties suitable for the needs of the modern society and those that promote changes using the abundant initiator of sunlight to trigger the changes are the most prominent candidates.

Acknowledgements

The authors thank to CNPq (grant 407619/2013-5) and FAPEG (grant 2012210267000923) for financial support.

Author details

Tatiana Duque Martins[1]*, Antonio Carlos Chaves Ribeiro[1], Geovany Albino de Souza[1], Diericon de Sousa Cordeiro[1], Ramon Miranda Silva[1], Flavio Colmati[1], Roberto Batista de Lima[2], Lucas Fernandes Aguiar[1], Leandro Lima Carvalho[1], Renan Gustavo Coelho S. dos Reis[1] and Wemerson Daniel C. dos Santos[2]

*Address all correspondence to: tati.duque@gmail.com

1 Chemistry Institute, Federal University of Goiás, Goiânia, Brazil

2 Chemistry Institute, Federal University of Maranhão, São Luís, Brazil

References

[1] Turro NJ, Ramamurthy V, Scaiano JC. Modern Molecular Photochemistry of Organic Molecules. 1st ed. USA: University Science Books; 2010. p. 1084

[2] Hollas JM. Modern Spectroscopy. 4th ed. England: John Wiley & Sons Ltd; 2004. p. 452

[3] Lewis GN, Kasha M. Phosphorescence and the triplet state. Journal of the American Chemical Society. 1944;**66**:2100-2116

[4] Lakowicz JR. Principles of Fluorescence Spectroscopy. 3rd ed. USA: Springer; 2006. p. 954

[5] Soboleva IV, van Stam J, Dutt GB, Kuzmin MG, De Schryver FC. Electron-transfer reactions in SDS micelles: Reactivity of pyrene and tris(2,2'-bipyridyl)ruthenium(II) excited states investigated by time-resolved luminescence quenching. Langmuir. 1999;**15**:6201-6207

[6] Rae M, Berberan-Santos MN. Pre-equilibrium approximation in chemical and photophysical kinetics. Chemical Physics. 2002;**280**:283-293

[7] Elbin M, Bazan GC. A new design strategy for organic optoelectronic materials by lateral boryl substitution. Angewandte Chemie International Edition. 2008;**47**:834-838

[8] Wang K, Murahari P, Yokoyama K, Lord JS, Pratt FL, He J, Schulz L, Willis M, Anthony JE, Morley NA, Nuccio N, Misquitta A, Dunstan DJ, Shimomura K, Watanabe I, Zhang S, Heathcote P, Drew AJ. Temporal mapping of photochemical reactions and molecular excited states with carbon specificity. Nature Materials. 2017;**16**:467-474

[9] Fresnadillo DG, Marazuela MD, Moreno-Bondi MC, Orellana G. Luminescent Nafion membranes dyed with ruthenium(II) complexes as sensing materials for dissolved oxygen. Langmuir. 1999;**15**:6451-6459

[10] Baur JW, Rubner MF, Reynolds JR, Kim S. Forster energy transfer studies of polyelectrolyte heterostructures containing conjugated polymers: A means to estimate layer interpenetration. Langmuir. 1999;**15**:6460-6469

[11] Mitsui M, Fukui H, Takahashi H, Takakura Y, Mizukami T. Single-molecule fluorescence spectroscopy of perylene diimide dyes in a γ-cyclodextrin film: Manifestation of

photoinduced H atom transfer via higher triplet (n, π*) excited states. The Journal of Physical Chemistry. A. 2017;**121**:1577-1586

[12] Arkan F, Izadyar M. The role of solvent and structure in the kinetics of the excitons in porphyrin-based hybrid solar cells. Solar Energy. 2017;146:368-378

[13] Kisslinger R, Hua W, Shankar K. Bulk heterojunction solar cells based on blends of conjugated polymers with II-VI and IV-VI inorganic semiconductor quantum dots. Polymers. 2017;**9**:35 (29 p). DOI: 10.3390/polym9020035

[14] Liu R. Hybrid organic/inorganic nanocomposites for photovoltaic cells. Materials. 2014;**7**: 2747-2771

[15] Dar MI, Franckevicius MM, Arora N, Redeckas K, Vengris M, Gulbinas V, Zakeeruddin SM, Grätzel M. High photovoltage in perovskite solar cells: New physical insights from the ultrafast transient absorption spectroscopy. Chemical Physics Letters. 2017;**683**:211-215. DOI: 10.1016/j.cplett.2017.04.046

[16] Tiedje T, Rose A. A physical interpretation of dispersive transport in disordered semiconductors. Solid State Communications. 1981;**37**:49-52

[17] Beard MC, Johnson JC, Luther JM, Nozik AJ. Multiple exciton generation in quantum dots versus singlet fission in molecular chromophores for solar photon conversion. Philosophical Transactions of the Royal Society A. 2015;**373**:20140412. DOI: 10.1098/rsta.2014.0412

[18] Beard MC, Luther JM, Semonin O, Nozik AJ. Third generation photovoltaics based on multiple exciton generation in quantum confined semiconductors. Accounts of Chemical Research. 2013;**46**:1252-1260. DOI: 10.1021/ar3001958

[19] Smith MB, Michl J. Singlet fission. Chemical Reviews. 2010;**110**:6891-6936. DOI: 10.1021/cr1002613

[20] Smith MB, Michl J. Recent advances in singlet fission. Annual Review of Physical Chemistry. 2013;**64**:361-386. DOI: 10.1146/annurev-physchem-040412-110130

[21] Gabor NM. Exciton multiplication in graphene. Accounts of Chemical Research. 2013;**46**: 1348-1357. DOI: 10.1021/ar300189j

[22] Thompson NJ, Congreve DN, Goldberg D, Menon VM, Baldo MA. Slow light enhanced singlet exciton fission solar cells with a 126% yield of electrons per photon. Applied Physics Letters. 2013;**103**:263302. DOI: 10.1063/1.4858176

[23] Semonin OE, Luther JM, Choi S, Chem HY, Gao J, Nozik AJ, Beard MC. Peak external quantum photocurrent efficiency exceeding 100% via MEG in a quantum dot solar cell. Science. 2011;**334**:1530-1533. DOI: 10.1126/science.1209845

[24] Wu B, Zhang G, Zhang S. Fate and implication of acetylacetone in photochemical processes for water treatment. Water Research. 2016;**101**:233-240

[25] Souza GA. Photophysical and Morphological Characterization of Peptide Nanostructures Containing Fluorescence Compound for Environmental Application. Goiânia/ Brazil. Federal University of Goiás; 2014

[26] Daud WRW, Rosli RE, Majlan EH, Hamid SAA, Mohamed R, Husaini T. PEM fuel cell system control: A review. Renewable Energy. 2017;**113**:620-638

[27] Baroutaji A, Carton JG, Stokes J, Olabi AG. Application of open pore cellular foam for air breathing PEM fuel cell. International Journal of Hydrogen Energy. 2017. in press, corrected proof. Available online 7 June 2017. DOI: 10.1016/j.ijhydene.2017.05.114

[28] Mench MM. Fuel Cell Engines. 1st ed. USA: John Wiley & Sons Inc.; 2008

[29] Mao L, Jackson L, Dunnett S. Electrolyte membrane (PEM) fuel cell system with data-driven approaches. Fuel Cells. 2017;**17**:247-258

[30] Setzler BP, Zhuang Z, Wittkopf JA, Yan Y. Activity targets for nanostructured platinum-group-metal-free catalysts in hydroxide exchange membrane fuel cells. Nature Nanotechnology. 2016;**11**:1020-1025

[31] Akhairi MAF, Kamarudin SK. Catalysts in direct ethanol fuel cell (DEFC): An overview. International Journal of Hydrogen Energy. 2016;**41**:4214-4228

[32] Wang Q, Sun GQ, Jiang LH, Xin Q, Sun SG, Jiang YX, Chen SP, Jusys Z, Behm RJ. Adsorption and oxidation of ethanol on colloid-based Pt/C, PtRu/C and Pt3Sn/C catalysts: In situ FTIR spectroscopy and on-line DEMS studies. Physical Chemistry Chemical Physics. 2007;**9**:2686-2696

[33] Rizo R, Sebastián D, Rodríguez JL, Lázaro MJ, Pastor E. Influence of the nature of the carbon support on the activity of Pt/C catalysts for ethanol and carbon monoxide oxidation. Journal of Catalysis. 2017;**348**:22-28

[34] Cantane DA, Ambrosio WF, Chatenet M, Lima FHB. Electro-oxidation of ethanol on Pt/C, Rh/C, and Pt/Rh/C-based electrocatalysts investigated by on-line DEMS. Journal of Electroanalytical Chemistry. 2012;**681**:56-65

[35] Beyhan S, Léger J-M, Kadırgan F. In situ FTIR investigation of acetic acid electrooxidation on carbon supported Pt–Sn based trimetallic catalysts: Influence of the nature of the third metal. Applied Surface Science. 2014;**321**:426-431

[36] Rightmire RA, Rowland RL, Boos DL, Beals DL. Ethyl alcohol oxidation at platinum electrodes. Journal of the Electrochemical Society. 1964;**111**:242-247

[37] Hitmi H, Belgsir EM, Léger J-M, Lamy C, Lezna RO. A kinetic analysis of the electro-oxidation of ethanol at a platinum electrode in acid medium. Electrochimica Acta. 1994;**39**:407-415

[38] Podlovchenko BI, Petry OA, Frumkin AN, Lal H. The behaviour of a platinized-platinum electrode in solutions of alcohols containing more than one carbon atom, aldehydes and formic acid. Journal of Electroanalytical Chemistry. 1966;**11**:12-25

[39] Antoniassi RM, Silva JCM, Oliveira-Neto A, Spinacé EV. Synthesis of Pt+Sno2/C electrocatalysts containing Pt nanoparticles with preferential (1 0 0) orientation for direct

ethanol fuel cell. Applied Catalysis B: Environmental. 2017. in press. DOI: 10.1016/j.apcatb.2017.06.031

[40] Lima FHB, Gonzalez ER. Ethanol electro-oxidation on carbon-supported Pt–Ru, Pt–Rh and Pt–Ru–Rh nanoparticles. Electrochimica Acta. 2008;**53**:2963-2971

[41] Santiago EI, Camara GA, Ticianelli EA. CO tolerance on PtMo/C electrocatalysts prepared by the formic acid method. Electrochimica Acta. 2003;**48**:3527-3534

[42] Kadirgan F, Beyhan S, Atilan T (2009) Preparation and characterization of nano-sized Pt–Pd/C catalysts and comparison of their electro-activity toward methanol and ethanol oxidation, Inter J Hydr Energy. **34**:4312–4320

[43] Lamy C. From hydrogen production by water electrolysis to its utilization in a PEM fuel cell or in a SO fuel cell: Some considerations on the energy efficiencies. International Journal of Hydrogen Energy. 2016;**41**:15415-15425

[44] Antolini E. Catalysts for direct ethanol fuel cells. Journal of Power Sources. 2017;**170**:1-12

[45] Parreira LS, Silva JCM, Simões FR, Cordeiro MAL, Sato RH, Leite ER, Santos MC. PtSn electrocatalyst supported on MWCNT-COOH: Investigating the ethanol oxidation reaction. ChemElectroChem. 2017;**4**:1-10

Chapter 5

Competition Kinetics: An Experimental Approach

Murtaza Sayed, Luqman Ali Shah, Javed Ali Khan,
Noor S. Shah, Rozina Khattak and Hasan M. Khan

Additional information is available at the end of the chapter

http://dx.doi.org/10.5772/intechopen.70483

Abstract

In this chapter, free radical kinetics with the help of competition kinetics and some experimental results calculated by competition kinetics to find out the rate constant of reactive species ($^{\bullet}OH$, e_{aq}^{-}, $^{\bullet}H$) with target compound, which is used by radiation chemists is briefly discussed. The competition kinetics method is well validated by taking ciprofloxacin, norfloxacin and bezafibrate as example compounds. The bimolecular rate constants of hydroxyl radical, hydrate electron and hydrogen atom has been calculated for example solute species (ciprofloxacin, norfloxacin and bezafibrate).

Keywords: competition kinetics, rate constants, norfloxacin, ciprofloxacin, bezafibrate

1. Introduction

Radiation chemistry involves extensive study of competition between fast reactions of transient species, reactive and intermediates. Such knowledge is useful to investigate the mechanism of a radiolytic reaction and to propose which process is taking place and which experimental condition is governing a reaction and to know the chemical kinetics of a radiolytic reaction under study.

Generally, in a chemical process, the reactant is converted to products in an individual step. However, in a radiation induced chemical reaction, all steps are taken into consideration including deposition of energy by a charged particle in the system and then formation of a final stable chemical product, and certainly will be a rather complex set of reactions [1]. In the following sections, we will briefly discuss the fast kinetics, i.e. competition kinetics to find the unknown rate constants of a compound with reactive species like hydroxyl radical ($^{\bullet}OH$) or hydrated electron (e_{aq}^{-}), by considering a reference compound whose rate constant with these reactive species is already known.

2. Competition kinetics

For detailed investigation of competing reactions, it is necessary to have a good knowledge about the rate constant data that will used to propose which reaction is predominant. For instance, the Fricke dosimeter contains three main active species (350 mol m^{-3} H^+, 1 mol m^{-3} Fe^{2+} and 0.25 mol m^{-3} O_2) that have very high rate constants with e_{aq}^- and to find out the reaction mechanism involved in the dosimetry, it is necessary to investigate which solute(s) will mainly react with e_{aq}^-.

The reactions and their corresponding rate constants are given as [1–3]:

$$e_{aq}^- + H^+ \rightarrow H \qquad k_1 = 2.3 \times 10^{10}\ M^{-1}s^{-1} \tag{1}$$

$$e_{aq}^- + Fe^{2+} \rightarrow Fe^+ \qquad k_2 = 1.6 \times 10^{8}\ M^{-1}s^{-1} \tag{2}$$

$$e_{aq}^- + O_2 \rightarrow O_2^- \qquad k_3 = 1.9 \times 10^{10}\ M^{-1}s^{-1} \tag{3}$$

The extent of reaction is proportional to the product k[solute] for each of the three solutes, that is,

$$e_{aq}^- + H^+ : e_{aq}^- + Fe^{2+} : e_{aq}^- + O_2 = k_1[H^+] : k_2\left[Fe^{2+}\right] : k_3[O_2] = 5 \times 10^4 : 1 : 30 \tag{4}$$

Alternatively, it can also be concluded that 99.94% of the hydrated electrons reacting with the three solutes will react with H^+, so under such conditions the reaction of e_{aq}^- with Fe^{2+} and O_2 will be ignored. Therefore, it is compulsory to have an wide collection of rate constant data to apply kinetics for a radiation induced chemical reaction.

In Fricke dosimeter, hydrogen ions are considered as strong scavengers of hydrated electrons [4–6]. The effectiveness of a chemical scavenger depends upon the product k[scavenger] that must have a higher value than k[substrate]. For example, *tert*-butanol is used to scavenge hydroxyl radicals and by using concentration of 1 mol m^{-3} *tert*-butanol it has k[tert-butanol] = $6.0 \times 10^5\ s^{-1}$. Therefore, *tert*-butanol would be an efficient scavenger for hydroxyl radical (can scavenger over 99% of the hydroxyl radical), if k[solute] for the reaction of hydroxyl radical with the solute is less than $6.0 \times 10^5\ s^{-1}$ [7]. Similarly, oxygen is used to scavenge hydrated electrons and hydrogen atoms from aqueous media [5, 8, 9] and for both the radical (e_{aq}^- and $^\bullet H$) in air-saturated media, $k[O_2] = 5 \times 10^6\ s^{-1}$, so that oxygen can be expected to interfere in the radiolysis of aqueous media if k[solute] for the reactions of hydrogen atoms and hydrated electrons with the solute are of the same order as, or less than, $5 \times 10^6\ s^{-1}$.

In case of radiolysis of organic species, their products also itself often act as scavengers and it is commonly found that the product yield is not in direct relation with the absorbed dose. To estimate the possible reasons of such effects, competition kinetics can be employed in an effective way if the radiolysis mechanism is known and the necessary rate constants are available. For example, cyclohexene is produced when cyclohexane is irradiated and both cyclohexene and cyclohexane have appreciable rate constants with hydrogen atoms, one of the radical specie produced during gamma radiolysis of aqueous media. The reactions are summarized below as:

$$^{\bullet}H + C_6H_{12} \rightarrow H_2 + {}^{\bullet}C_6H_{11} \qquad k_4 = 3.0 \times 10^7\ M^{-1}s^{-1} \tag{5}$$

$$^{\bullet}H + C_6H_{10} \rightarrow {}^{\bullet}C_6H_{11} \qquad k_5 = 3.0 \times 10^9\ M^{-1}s^{-1} \tag{6}$$

Under such conditions, the hydrogen atoms will be reacting equally with cyclohexane and cyclohexene when k_4 [cyclohexane] = k_5[cyclohexene] or in other words we can say that only when 1% of the cyclohexane has been converted to cyclohexene.

Furthermore, competition kinetics is also employed when measuring rate constants by pulse radiolysis using a reference compound.

2.1. Computation of bimolecular rate constant of $^{\bullet}$OH with ciprofloxacin

Ciprofloxacin (CIP) belongs to a class of fluoroquinolone family and is used globally as a human and veterinary medication [10]. It has been very much concentrated that the event of these wide range antibiotics in the water bodies may position genuine dangers to the environment and human wellbeing by producing expansion of bacterial medication inactivation. The natural event of these fluoroquinolones anti-infection agents in numerous nations, similar to Switzerland, Australia and China have been affirmed in recent literature [10–13]. It has likewise been watched that most quinolone antibiotics are not completely utilized in the human body and accordingly are discharged and acquainted with the amphibian condition through wastewater sewages because of poor execution of ordinary water treatment plants [12, 14–16] bringing about adversative impacts to sea-going microorganisms and fish [17, 18]. Thus, it becomes necessary to advice alternative physiochemical techniques for effective removal of these contaminants and diminish their ecological effects [12, 19]. For this reason, the deterioration of ciprofloxacin (CIP) in water utilizing ionizing radiations was evaluated to examine the rate constant of $^{\bullet}$OH with CIP. The degradation curves of CIP by gamma irradiation at various absorbed doses has been shown by **Figure 1**.

For computation of bimolecular rate constant of $^{\bullet}$OH with CIP by competition kinetics, phenol was selected as reference compound that has second order rate constant of $6.6 \times 10^9\ M^{-1}\ s^{-1}$ with $^{\bullet}$OH [20]. The sample solution of total 150 mL having CIP and reference compound phenol together in equivalent quantity was immersed with oxygen gas to change over e_{aq}^{-} and $^{\bullet}$H to superoxide radical anions quickly [21], which are less responsive compared to $^{\bullet}$OH.

$$CIP + \bullet OH \xrightarrow{k_{CIP}} \text{Products} \tag{7}$$

$$Phenol + \bullet OH \xrightarrow{k_{phenol}} \text{Products} \tag{8}$$

While, the notations, k_{CIP} and k_{phenol} denotes the bimolecular rate constants of $^{\bullet}$OH with CIP and phenol, respectively. Keeping the condition that the dose rate (DR) was kept constant, the rate of decay of CIP is directly related to the rate constant and the concentration of $^{\bullet}$OH [21]:

So,

$$-\frac{d[CIP]}{dD} = -\frac{1}{DR}\frac{d[CIP]}{dt} \tag{9}$$

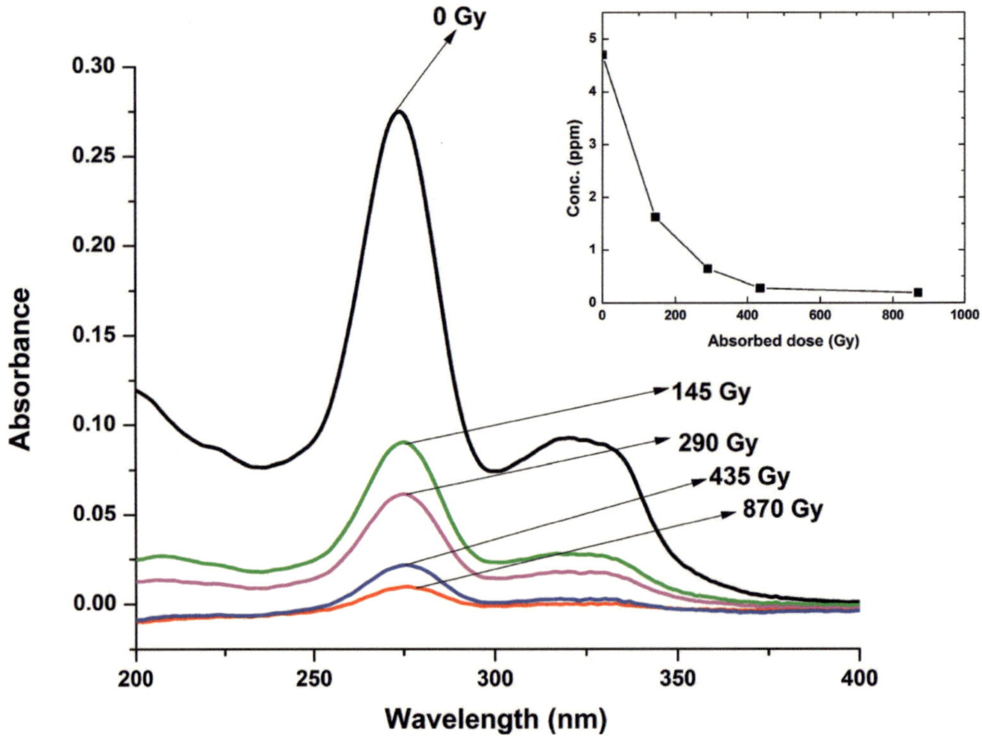

Figure 1. The UV spectra of CIP solution observed by gamma irradiation at various absorbed doses ranging from 0 to 870 Gy. Inset shows the influence of gamma-irradiation on degradation of 4.6 mg L^{-1} of CIP solution [8].

Or,

$$-\frac{d}{dD}[CIP] = -\frac{1}{DR}k_{CIP}[CIP][^{\bullet}\mathrm{OH}] \tag{10}$$

Similarly,

$$-\frac{d}{dD}[Phenol] = -\frac{1}{DR}k_{phenol}[Phenol][^{\bullet}\mathrm{OH}] \tag{11}$$

While the absorbed ionizing dose and the total time for which irradiation was performed, are represented by "D" and "t," respectively. Subsequently, the original concentrations of both CIP and phenol are same. Therefore, the rate of decay of CIP to phenol would be equal to the ratio of their individual rate constants as follows [8]:

$$\frac{-\frac{d}{dD}[CIP]}{-\frac{d}{dD}[phenol]} = \frac{k_{CIP}}{k_{phenol}} \tag{12}$$

Or,

$$\ln\frac{[CIP]_D}{[CIP]_0} = \frac{k_{CIP}}{k_{phenol}}\ln\frac{[phenol]_D}{[phenol]_0} \tag{13}$$

In Eq. (13), at time 0, the concentration of CIP and phenol are represented by $[CIP]_0$ and $[Phenol]_0$, respectively; while after absorbed dose "D" of gamma irradiation, the corresponding concentration of CIP and phenol are represented by $[CIP]_D$ and $[Phenol]_D$, respectively.

When a straight line is plotted by taking $\ln \frac{[CIP]_D}{[CIP]_0}$ vs $\ln \frac{[Phenol]_D}{[Phenol]_0}$ a slope with k_{CIP}/k_{Phenol} = 0.4012 was obtained, so we have;

$$0.4012 = \frac{k_{CIP}}{k_{Phenol}} \tag{14}$$

Or,

$$k_{CIP} = 2.64 \times 10^9 \ M^{-1}s^{-1}. \tag{15}$$

Henceforth, the bimolecular rate constant of $^{\bullet}OH$ with CIP was calculated to be $2.75 \times 10^9 \ M^{-1} \ s^{-1}$. Dodd et al. [22] likewise ascertained apparent second order rate constant of with $^{\bullet}OH$-radical with CIP to be 4.1 (±0.3) × $10^9 \ M^{-1} \ s^{-1}$, which is somewhat higher than the esteem calculated in the current report [8].

2.2. Computation of $^{\bullet}OH$, e_{aq}^- and $^{\bullet}H$ with norfloxacin

Norfloxacin (NORO) may be called likewise chemotherapeutic antibacterial agent, furthermore is regularly utilized to medicine for urinary tract infections [23]. Its occurrence in surface water and wastewater overflows has been accounted for at follow ppb levels [12, 14, 24–27]. Even though, the detected concentration of NORO is very low and normally ranges from ng L^{-1} to μg L^{-1} in water bodies and μg kg^{-1} to mg kg^{-1} in soils and sediments, still these fluoroquinolone family are categorized as "pseudopersistant" contaminants because of their continuous and regular discharge into the water bodies [28, 29]. González-Pleiter et al. [30] concentrated on those unique united toxicities from claiming norfloxacin, amoxicillin, erythromycin, levofloxacin, furthermore anti-microbial prescription toward two maritime organisms, i.e. Cyanobacterium Anabaena CPB 4337. Similarly as a goal existing being and the green alga *Pseudokirchneriella subcapitata* as a non-target existing continuously. They assigned norfloxacin on a chance to be a greater amount dangerous on cyanobacterium over green alga. Furthermore, norfloxacin alone and additionally its mixture for different antibiotics might stance genuine idle danger to oceanic environment.

The presence of NORO in the fresh water bodies indicate that traditional wastewater or water treatment techniques are not efficient to remove NORO from aquatic environment due to its aromatic nature and its occurrence cause thoughtful health associated problems by using contaminated drinking water [31–33]. Therefore, it becomes an issue of interest to remove NORO from the aquatic environment.

In a typical experiment for gamma radiolysis of NORO, the apparent bimolecular rate constant of $^{\bullet}OH$, e_{aq}^- and $^{\bullet}H$ with NORO was assessed, using competition kinetics.

The following Eq. (14) was employed to measure the bimolecular rate constant of $^{\bullet}OH$, e_{aq}^- and $^{\bullet}H$, which are the main species produced during gamma radiolysis of aqueous media [21].

$$k_{^\bullet OH/NORO} = \frac{\ln\left([NORO]_0/[NORO]_D\right)}{\ln\left([2-CP]_0/[2-CP]_D\right)} k_{^\bullet OH/2-CP} \tag{16}$$

2-Chlorophenol (2-CP) was selected as reference compound which have recognized rate constants with $^\bullet$OH, e_{aq}^- and $^\bullet$H ($k_{\bullet OH/2\text{-}CP} = 1.2 \times 10^{10}\ M^{-1}\ s^{-1}$, $k_{e_{aq}^-/2-CP} = 1.3 \times 10^9\ M^{-1}s^{-1}$, $k_{\bullet H/2\text{-}CP} = 1.5 \times 10^9\ M^{-1}\ s^{-1}$) [20]. To permit only $^\bullet$OH to react with NORO, and scavenge e_{aq}^- and $^\bullet$H, O_2 saturated sample (O_2 changes e_{aq}^- and $^\bullet$H to superoxide radical anions, which are less responsive opposite to $^\bullet$OH) [20] was applied for computing the bimolecular rate of $^\bullet$OH with NORO. In the same way, the bimolecular rate constant of e_{aq}^- with NORO (ke_{aq}–/NORO) was calculated by N_2-puging the sample solution added with 0.1 M *iso*-propanol (*iso*-propanol is used to scavenges both $^\bullet$OH and $^\bullet$H) [20]. Similarly, the computation of bimolecular constant of $^\bullet$H with NORO ($k_{\bullet H/NORO}$) was made by N_2 saturating the solution of 0.1 M *tert*-butanol (*tert*-butanol is used to scavenge $^\bullet$OH) [20] at pH 2.2. Low pH was maintained to get high yield of $^\bullet$H through reaction of e_{aq}^- with $^+$H [34].

A linear plot with slope equal to $k_{\bullet OH/NORO}/k_{\bullet OH/2\text{-}CP}$ was observed by plotting ln([NORO]$_0$/ [NORO]$_D$) vs ln ([2-CP]$_0$/[2-CP]$_D$) at several absorbed ionizing doses. The same calculation was implemented for measurement of bimolecular rate constant of e_{aq}^- and $^\bullet$H with NORO, respectively. Applying the obtained slope values, the second order rate constants of $^\bullet$OH, e_{aq}^- and $^\bullet$H with NORO were computed to be $(8.81 \pm 0.03) \times 10^9\ M^{-1}\ s^{-1}$, $(9.54 \pm 0.16) \times 10^8\ M^{-1}\ s^{-1}$ and $(1.10 \pm 0.20) \times 10^9\ M^{-1}\ s^{-1}$, respectively [5], which also indicates that $k_{e_{aq}^-/NORO}$ is lesser to $k_{\bullet H/NORO,}$ or in other words the reactivity of e_{aq}^- to NORO is less than the reactivity of $^\bullet$H with NORO. Thus, in the removal of NORO by ionizing irradiation $^\bullet$H is of immense importance. The bimolecular rate constant of $^\bullet$OH with NORO in the current report is analogous with the study of Santoke et al. [35], in which they calculated the bimolecular rate constants of $^\bullet$OH with six common fluoroquinolones (orbifloxacin, flumequine, marbofloxacin, danofloxacin, enrofloxacin and model compound, 6-fluoro-4-oxo-1,4-dihydro-3-quinolone carboxylic acids) and was found to be in the range of $6.4 - 9.03 \times 10^9\ M^{-1}\ s^{-1}$.

2.3. Measurement of bimolecular rate constant of $^\bullet$OH with bezafibrate

Bezafibrate (BZF) is also the most commonly detected pollutant among various pharmaceuticals excreted into the sewage system and is categorized as persistent organic pollutants [36]. In drinking water its concentration has been noticed at the levels of 27 ng L^{-1} [37] in rivers at the concentrations level of 0.1–0.15 μg L^{-1} [37], in small streams in the range of 0.5–1.9 μg L^{-1} [37], in surface waters in the range of 3.1 μg L^{-1} [38], and up to 4.6 μg L^{-1} level in sewage treatment plant effluents. Owing to its high use and persistence nature, the elimination of BZF from aqueous media has emerged as a hot research topic. The qualitative and quantitative analysis of its degradation products besides its degradation kinetics is also of great concern. Keeping in view all these problems, the degradation of BZF was investigated by photo catalysis using hydrothermally synthesized TiO_2/Ti films with exposed {001} facets. Besides photo catalysis, there are other many advanced treatment options for efficient removal of BZF from aqueous media, such as nanofiltration techniques, ultraviolet (UV) radiation and advanced oxidation processes (AOPs) [39] and these have been thoroughly studied. In AOPs (the most reliable and efficient technique), as compared to other treatment techniques the pollutant of interest is converted in to more stable, harmless inorganic species such as carbon dioxide, water and mineral salts. AOPs are categorized as ozonation (O_3), H_2O_2, O_3/H_2O_2/photocatalysis, and O_3/H_2O_2/UV photocatalysis [39, 40]. TiO_2 photocatalysis is considered as more auspicious and efficient technique among semiconductor photocatalysis [41, 42]. TiO_2 photo active material has shown a great potential in many applications, including water splitting to generate O_2 and H_2 [43, 44] water and wastewater treatment [45, 46], gas phase treatment [47, 48], as well as in solar cells [49]. So, in this case the degradation of BZF was performed by VUV photo active material with exposed {001} faceted TiO_2/Ti material.

For measurement of absolute bimolecular rate constant of •OH with BZF, *para*-chlorobenzoic acid (*p*-CBA) was used as probe molecule was calculated using *para*-chlorobenzoic acid (*p*-CBA) as probe molecule and by employing competition kinetics technique established by Pereira et al. [50] and given in Eq. (15).

$$k_{\bullet OH(s)} = \frac{k_{s(UV/H_2O_2)} - k_{s(UV)}}{k_{ref(UV/H_2O_2)} - k_{ref}} \times k_{\bullet OH(ref)} \tag{17}$$

Where, $k_{\bullet OH}$, $k_{(UV/H_2O_2)}$ and $k_{(UV)}$ represent the second order rate constant of hydroxyl radical, UV/H_2O_2, fluence based rate constant of UV/hydrogen peroxide process and UV, fluence based rate constant of direct photolysis, respectively. The notations "s" and "ref" represents the substrate and reference compounds, which in our case is BZF and *p*-CBA, respectively. For the determination of $^{\bullet}OH$ rate constant with BZF, two sets of experiments were performed. In one set of experiments, the solution containing 27.63 μM of BZF, 27.63 μM of *p*-CBA and 1 mM

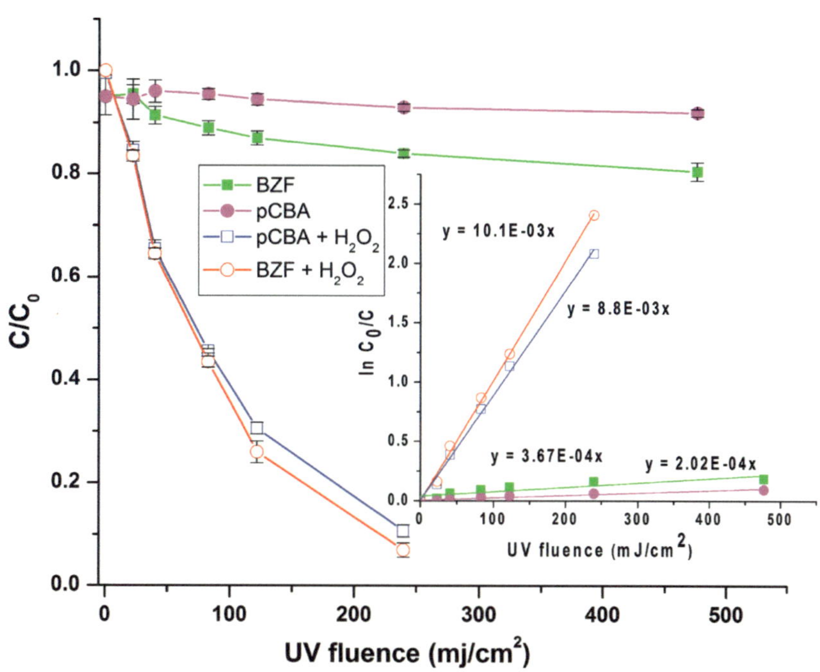

Figure 2. Determination of bimolecular rate constant of BZF with $^{\bullet}OH$; inset shows the degradation kinetics of BZF alone, p-CBA alone, BZF+ H_2O_2, pCBA + H_2O_2 exposed to UV-irradiation.

of H_2O_2 was exposed to UV irradiation, while another set of experiments was free of H_2O_2 to calculate k_{UV}. The concentration of H_2O_2 was kept higher for ensuring production of efficient $^{\bullet}OH$ with UV-photocatalysis.

Figure 2 shows degradation curves for BZF and *p*-CBA, both BZF and *p*-CBA were found to follow pseudo-first order degradation kinetics. The second order rate constant of $^{\bullet}OH$ with p-CBA is $5.0 \times 10^9 \ M^{-1} \ s^{-1}$ [20]. By substituting pseudo-first order degradation constants (k_{UV/H_2O_2} and k_{UV}) values in Eq. (15), the bimolecular rate constant of $^{\bullet}OH$ with BZF was calculated and found out to be $5.66 \times 10^9 \ M^{-1} \ s^{-1}$.

3. Conclusions

The overall conclusion of this chapter is that, in radiation chemistry to have a good knowledge about the mechanism of a reaction mechanism it is necessary that one must have sufficient understanding about the free radical kinetics. In addition, competition kinetics model can be successful applied for the determination of unknown rate constants of reactive species with solute molecule. The competition kinetics can not only be applied for $^{\bullet}OH$ rate constants with the solute but also for measurement of e_{aq}^{-} and $^{\bullet}H$ with the target species. The competition kinetics method is validated by taking ciprofloxacin, norfloxacin and bezafibrate as example compounds. However, it should be make sure that competing reactions do not disobey the kinetics rules.

Author details

Murtaza Sayed[1]*, Luqman Ali Shah[1], Javed Ali Khan[1], Noor S. Shah[1,2], Rozina Khattak[3] and Hasan M. Khan[1]

*Address all correspondence to: murtazasayed_407@yahoo.com

1 Radiation and Environmental Chemistry Laboratory, National Centre of Excellence in Physical Chemistry, University of Peshawar, Pakistan

2 Department of Environmental Sciences, COMSATS Institute of Information Technology, Vehari, Pakistan

3 Department of Chemistry, Shaheed Benazir Bhutto Women University, Peshawar, Pakistan

References

[1] Spinks JWT, Woods RJ. An Introduction to Radiation Chemistry. John-Wiley and Sons, Inc., New York; 1990.

[2] Fricke H, Hart EJ. Chemical dosimetry. In: Radiation Dosimetry 2. Academic Press, New York; 1966. pp. 167-239

[3] Buxton GV, Stuart CR. Re-evaluation of the thiocyanate dosimeter for pulse radiolysis. Journal of the Chemical Society, Faraday Transactions 1995;**91**:279-281

[4] Sayed M, Fu P, Shah LA, Khan HM, Nisar J, Ismail M, Zhang P. VUV-photocatalytic degradation of bezafibrate by hydrothermally synthesized enhanced {001} facets TiO2/Ti film. The Journal of Physical Chemistry A 2015;**120**:118-127

[5] Sayed M, Khan JA, Shah LA, Shah NS, Khan HM, Rehman F, Khan AR, Khan AM. Degradation of quinolone antibiotic, norfloxacin, in aqueous solution using gamma-ray irradiation. Environmental Science and Pollution Research 2016;**23**:13155-13168

[6] Sayed M, Shah LA, Khan JA, Shah NS, Khan HM, Khan RA, Khan AR, Khan AM. Hydroxyl radical based degradation of ciprofloxacin in aqueous solution. Journal of the Chilean Chemical Society 2016;**61**:2949-2953

[7] Rehman F, Murtaza S, Ali Khan J, Khan HM. Removal of crystal violet dye from aqueous solution by gamma irradiation. Journal of the Chilean Chemical Society 2017;**62**:3359-3364

[8] Sayed M, Ismail M, Khan S, Tabassum S, Khan HM. Degradation of ciprofloxacin in water by advanced oxidation process: Kinetics study, influencing parameters and degradation pathways. Environmental Technology 2016;**37**:590-602

[9] Nisar J, Sayed M, Khan FU, Khan HM, Iqbal M, Khan RA, Anas M. Gamma-irradiation induced degradation of diclofenac in aqueous solution: Kinetics, role of reactive species and influence of natural water parameters. Journal of Environmental Chemical Engineering 2016;**4**:2573-2584

[10] Zhang H, Huang C-H. Oxidative transformation of fluoroquinolone antibacterial agents and structurally related amines by manganese oxide. Environmental Science & Technology 2005;**39**:4474-4483

[11] Ikehata K, Gamal El-Din M, Snyder SA. Ozonation and advanced oxidation treatment of emerging organic pollutants in water and wastewater. Ozone: Science & Engineering 2008;**30**:21-26

[12] Watkinson AJ, Murby EJ, Costanzo SD. Removal of antibiotics in conventional and advanced wastewater treatment: Implications for environmental discharge and wastewater recycling. Water Research 2007;**41**:4164-4176

[13] Xu W, Zhang G, Li X, Zou S, Li P, Hu Z, Li J. Occurrence and elimination of antibiotics at four sewage treatment plants in the Pearl River Delta (PRD), South China. Water Research 2007;**41**:4526-4534

[14] Kümmerer K, Al-Ahmad A, Mersch-Sundermann V. Biodegradability of some antibiotics, elimination of the genotoxicity and affection of wastewater bacteria in a simple test. Chemosphere 2000;**40**:701-710

[15] Mascarelli AL. New mode of action found for pharmaceuticals in the environment. Environmental Science & Technology 2010;**44**:1159-1160

[16] Kümmerer K. Antibiotics in the aquatic environment—a review—part I. Chemosphere 2009;**75**:417-434

[17] Christensen AM, Ingerslev F, Baun A. Ecotoxicity of mixtures of antibiotics used in aquacultures. Environmental Toxicology and Chemistry 2006;**25**:2208-2215

[18] Carlsson G, Orn S, Larsson D. Effluent from bulk drug production is toxic to aquatic vertebrates. Environmental Toxicology and Chemistry 2009;**28**:2656-2662

[19] Andreozzi R, Campanella L, Fraysse B, Garric J, Gonnella A, Giudice RL, Marotta R, Pinto G, Pollio A. Effects of advanced oxidation processes (AOPs) on the toxicity of a mixture of pharmaceuticals. Water Science and Technology 2004;**50**:23-28

[20] Buxton G, Greenstock C, Helman W, Ross A. Critical review of rate constants for reactions of hydrated electrons, hydrogen-atoms and hydroxyl radicals $^{\cdot}OH/^{\cdot}O^{-}$ in aqueous solution. Journal of Physical and Chemical Reference Data 1988;**17**:513-886

[21] Kimura A, Osawa M, Taguchi M. Decomposition of persistent pharmaceuticals in wastewater by ionizing radiation. Radiation Physics and Chemistry 2012;**81**:1508-1512

[22] Dodd MC, Buffle M-O, Von Gunten U. Oxidation of antibacterial molecules by aqueous ozone: Moiety-specific reaction kinetics and application to ozone-based wastewater treatment. Environmental Science & Technology 2006;**40**:1969-1977

[23] Prieto A, Moder M, Rodil R, Adrian L, Marco-Urrea E. Degradation of the antibiotics norfloxacin and ciprofloxacin by a white-rot fungus and identification of degradation products. Bioresource Technology 2011;**102**:10987-10995

[24] Kolpin DW, Furlong ET, Meyer MT, Thurman EM, Zaugg SD, Barber LB, Buxton HT. Pharmaceuticals, hormones, and other organic wastewater contaminants in US streams, 1999-2000: A national reconnaissance. Environmental Science & Technology 2002;**36**:1202-1211

[25] Leung HW, Minh TB, Murphy MB, Lam JC, So MK, Martin M, Lam PK, Richardson BJ. Distribution, fate and risk assessment of antibiotics in sewage treatment plants in Hong Kong, South China. Environment International 2012;**42**:1-9

[26] Gracia-Lor E, Sancho JV, Serrano R, Hernandez F. Occurrence and removal of pharmaceuticals in wastewater treatment plants at the Spanish Mediterranean area of Valencia. Chemosphere 2012;**87**:453-462

[27] Oberlé K, Capdeville M-J, Berthe T, Budzinski Hln, Petit F. Evidence for a complex relationship between antibiotics and antibiotic-resistant *Escherichia coli*: From medical center patients to a receiving environment. Environmental Science & Technology. 2012; **46**:1859-1868

[28] Lindberg RH, Olofsson U, Rendahl P, Johansson MI, Tysklind M, Andersson BAV. Behavior of fluoroquinolones and trimethoprim during mechanical, chemical, and active sludge treatment of sewage water and digestion of sludge. Environmental Science & Technology 2006;**40**:1042-1048

[29] Hernando MD, Mezcua M, Fernandez-Alba AR, Barcelo D. Environmental risk assessment of pharmaceutical residues in wastewater effluents, surface waters and sediments. Talanta 2006;**69**:334-342

[30] González-Pleiter M, Gonzalo S, Rodea-Palomares I, Leganés F, Rosal R, Boltes K, Marco E, Fernández-Piñas F. Toxicity of five antibiotics and their mixtures towards photosynthetic aquatic organisms: Implications for environmental risk assessment. Water Research 2013;**47**:2050-2064

[31] Liu WF, Zhang J, Zhang CL, Ren L. Sorption of norfloxacin by lotus stalk-based activated carbon and iron-doped activated alumina: Mechanisms, isotherms and kinetics. Chemical Engineering Journal 2011;**171**:431-438

[32] Rosal R, Rodriguez A, Perdigon-Melon JA, Petre A, Garcia-Calvo E, Gomez MJ, Aguera A, Fernandez-Alba AR. Occurrence of emerging pollutants in urban wastewater and their removal through biological treatment followed by ozonation. Water Research 2010;**44**: 578-588

[33] Xu WH, Zhang G, Zou SC, Li XD, Liu YC. Determination of selected antibiotics in the Victoria Harbour and the Pearl River, South China using high-performance liquid chromatography-electrospray ionization tandem mass spectrometry. Environmental Pollution 2007;**145**:672-679

[34] Sánchez-Polo M, López-Peñalver J, Prados-Joya G, Ferro-García MA, Rivera-Utrilla J. Gamma irradiation of pharmaceutical compounds, nitroimidazoles, as a new alternative for water treatment. Water Research 2009;**43**:4028-4036

[35] Santoke H, Song W, Cooper WJ, Greaves J, Miller GE. Free-radical-induced oxidative and reductive degradation of fluoroquinolone pharmaceuticals: Kinetic studies and degradation mechanism. The Journal of Physical Chemistry A 2009;**113**:7846-7851

[36] Sui Q, Gebhardt W, Schröder HF, Zhao W, Lu S, Yu G. Identification of new oxidation products of bezafibrate for better understanding of its toxicity evolution and oxidation mechanisms during ozonation. Environmental Science & Technology 2017;**51**:2262-2270

[37] Ternes T. Pharmaceuticals and Metabolites as Contaminants of the Aquatic Environment. ACS Publications; 2001.

[38] Weston A, Caminada D, Galicia H, Fent K. Effects of lipid-lowering pharmaceuticals bezafibrate and clofibric acid on lipid metabolism in fathead minnow (Pimephales promelas). Environmental Toxicology and Chemistry. American Chemical Society. 2009;**28**: 2648-2655

[39] Klavarioti M, Mantzavinos D, Kassinos D. Removal of residual pharmaceuticals from aqueous systems by advanced oxidation processes. Environment International 2009;**35**: 402-417

[40] Parsons S. Advanced Oxidation Processes for Water and Wastewater Treatment. IWA publishing; 2004

[41] Madhavan J, Grieser F, Ashokkumar M. Combined advanced oxidation processes for the synergistic degradation of ibuprofen in aqueous environments. Journal of Hazardous Materials. American Chemical Society. 2010;**178**:202-208

[42] Grabowska E, Diak M, Marchelek M., Zaleska A. Decahedral TiO_2 with exposed facets: Synthesis, properties, photoactivity and applications. Applied Catalysis B: Environmental 2014;**156**:213-235

[43] Kim HS, Kim D, Kwak BS, Han GB, Um M-H, Kang M. Synthesis of magnetically separable core@ shell structured NiFe2O4@TiO2 nanomaterial and its use for photocatalytic hydrogen production by methanol/water splitting. Chemical Engineering Journal 2014;**243**:272-279

[44] Wang C, Hu Q-Q, Huang J-Q, Deng Z-H, Shi H-L, Wu L, Liu Z-G, Cao Y-G. Effective water splitting using N-doped TiO2 films: Role of preferred orientation on hydrogen production. International Journal of Hydrogen Energy 2014;**39**:1967-1971

[45] Choi J, Lee H, Choi Y, Kim S, Lee S, Lee S, Choi W, Lee J. Heterogeneous photocatalytic treatment of pharmaceutical micropollutants: Effects of wastewater effluent matrix and catalyst modifications. Applied Catalysis B: Environmental 2014;**147**:8-16

[46] Lee HU, Lee G, Park JC, Lee Y-C, Lee SM, Son B, Park SY, Kim C, Lee S, Lee SC. Efficient visible-light responsive TiO_2 nanoparticles incorporated magnetic carbon photocatalysts. Chemical Engineering Journal 2014;**240**:91-98

[47] Mrowetz M, Balcerski W, Colussi A, Hoffmann MR. Oxidative power of nitrogen-doped TiO_2 photocatalysts under visible illumination. The Journal of Physical Chemistry B 2004; **108**:17269-17273

[48] Nischk M, Mazierski P, Gazda M, Zaleska A. Ordered TiO_2 nanotubes: The effect of preparation parameters on the photocatalytic activity in air purification process. Applied Catalysis B: Environmental 2014;**144**:674-685

[49] Dwivedi C, Dutta V, Chandiran AK, Nazeeruddin MK, Grätzel M. Anatase TiO_2 hollow microspheres fabricated by continuous spray pyrolysis as a scattering layer in dye-sensitised solar cells. Energy Procedia 2013;**33**:223-227

[50] Pereira VJ, Weinberg HS, Linden KG, Singer PC. UV degradation kinetics and modeling of pharmaceutical compounds in laboratory grade and surface water via direct and indirect photolysis at 254 nm. Environmental Science & Technology 2007;**41**:1682-1688

Chapter 6

Catalyst Kinetics and Stability in Homogeneous Alcohol Acceptorless Dehydrogenation

Martin Nielsen

Additional information is available at the end of the chapter

http://dx.doi.org/10.5772/intechopen.70654

Abstract

The anthropogenic climate changes caused by meeting the energy demands of society by use of fossil fuels render the development of benign alternatives imperative. Probably, the most promising alternative is generating energy by means of power units driven by, e.g., solar, wind, water, etc., and then storing the energy that is not immediately used in battery type devices. Such a device might consist of hydrogen chemically stored as alcohol(s). The advantage of this method is that it allows gaseous hydrogen to be stored much more efficiently when liquefied as an alcohol. Moreover, as will be shown in this review, it is possible to release the hydrogen under mild conditions when employing homogeneous catalysis. This review considers the kinetic aspects of homogeneously catalysed acceptorless alcohol dehydrogenation reactions. For clarity, the sections are divided according to alcohol substrate, and each metal are described and discussed in subsections. Moreover, the kinetic information in the homogeneously catalysed AAD is traditionally provided simply as the turnover frequency, and more in-depth studies on the actual kinetic parameters are to date still largely elusive.

Keywords: homogeneous catalysis, acceptorless dehydrogenation, alcohols, catalyst kinetics, catalyst stability

1. Introduction

Acceptorless alcohol dehydrogenation (AAD) is the extrusion of H_2 from alcohols resulting in H_2 and carbonyl products. The latter includes, e.g., aldehydes, ketones, esters, amides, carboxylic acids, and CO_2. Conducting AAD by means of homogeneous catalysis is a promising approach towards a viable applicable energy carrier technology. For example, the vision with "Methanol economy" [1] is that H_2 produced by renewable energy sources is stored and transported in MeOH. The reasons for doing so are that H_2 is unfeasible to transport and that renewables are intermittent energy sources, which is unfit with the continuous energy need from society. Another example is the direct use bioalcohols as H_2 sources. It can be envisioned that the H_2 can be stored directly in the renewables by using, e.g., bioethanol or glycerol as the resource materials.

Besides the energy application, AAD has demonstrated its usefulness in a plethora of preparative systems. This type of chemistry focuses mainly on the transformation of organic functional groups and will as such not be covered in this review.

Charman studied the fundamental principles of AAD with isopropanol as a model substrate and $[RhCl_6]^{3-}$ as a catalyst in the 1960s [2], and Robinson made further advances in the 1970s [3, 4]. As such, the field of AAD has been active for more than 5 decades. Nevertheless, it can be argued that the area is still immature and much fundamental research is still imperative to take the technology towards methods feasible for commercial application. This review aims to contribute to that end by shedding light on the kinetics and stabilities of various AAD systems mainly developed in the last approximately 10 years. For brevity, focus will be on contributions that provide both catalyst activity and longevity investigations on reactions using isopropanol, ethanol, or methanol.

2. Secondary alcohols

Secondary alcohols are notoriously easier to dehydrogenate than primary alcohols for several reasons. The resulting ketone from dehydrogenating a secondary alcohol is more stable than the corresponding aldehyde, both from a thermodynamic and kinetic perspective. In addition, the aldehyde may easily react further reaching more oxidised functional groups, such as ester, carboxylic acid, or amide depending on the reaction conditions.

2.1. Isopropanol

In 1967, Charman reported that a turnover frequency (TOF) of approximately 14 h^{-1} can be achieved by employing a mixture of 7.6×10^{-3} M (580 ppm) $RuCl_3$, 9.4×10^{-2} M LiCl and 5.5×10^{-2} M HCl in refluxing isopropanol [2]. A decade later, Robinson reported that a combination of 4.45×10^{-2} M (3400 ppm) of $[Ru(OCOCF_3)_2(CO)(PPh_3)_2]$ and 12 equivalent trifluoroacetic acid in refluxing isopropanol led to an initial TOF of approximately 13 h^{-1} [3, 4]. After an additional 10 more years, Cole-Hamilton demonstrated that a TOF of 330 h^{-1} could be reached by using 1.96×10^{-4} M (15 ppm) $RuH_2(N_2)(PPh_3)_3$ and 1 M NaOH at 150°C [5]. The Charman and Robinson systems employ acidic environments, whereas the Cole-Hamilton system is alkaline. However, a direct comparison between the systems and the effect of the additive is hampered by the large reaction temperature and catalyst loading differences, where Cole-Hamilton uses a reaction temperature highly elevated and considerably less concentrated catalyst compared to the others.

In general, the mechanisms were believed to involve an inner-sphere β-hydride elimination of the alcohol followed by proton-assisted H_2 extrusion from the organometallic catalytic intermediate. The proton source would be the acid when present (Charman and Robinson systems); otherwise, the alcohol itself served as the proton donor, which concurrently formed

http://dx.doi.org/10.5772/intechopen.70654

an activated alcoholate more prone for β-hydride elimination. Hence, a bimolecular transition state involving the transient catalyst intermediate and either the acid or alcohol is evoked.

Robinson observed that the catalytic efficacy was heavily depending on the stoichiometry of the trifluoroacetic acid, corroborating its involvement in the rate-determining catalytic step(s). **Figure 1** illustrates the main elemental steps of the proposed catalytic cycle. Commencing with complex **A**, isopropanol coordination leads to **B**, which is then ready for extrusion of trifluoroacetic acid. Upon this extrusion, the β-hydride elimination leads to an acetone coordinated complex **C**, and complex **D** is then formed by extruding the acetone molecule. Hydrogen formation from complex **D** occurs by trifluoroacetic acid mediated protonation of the hydride in **D**, which then also regenerates complex **A** and thereby closes the catalytic cycle.

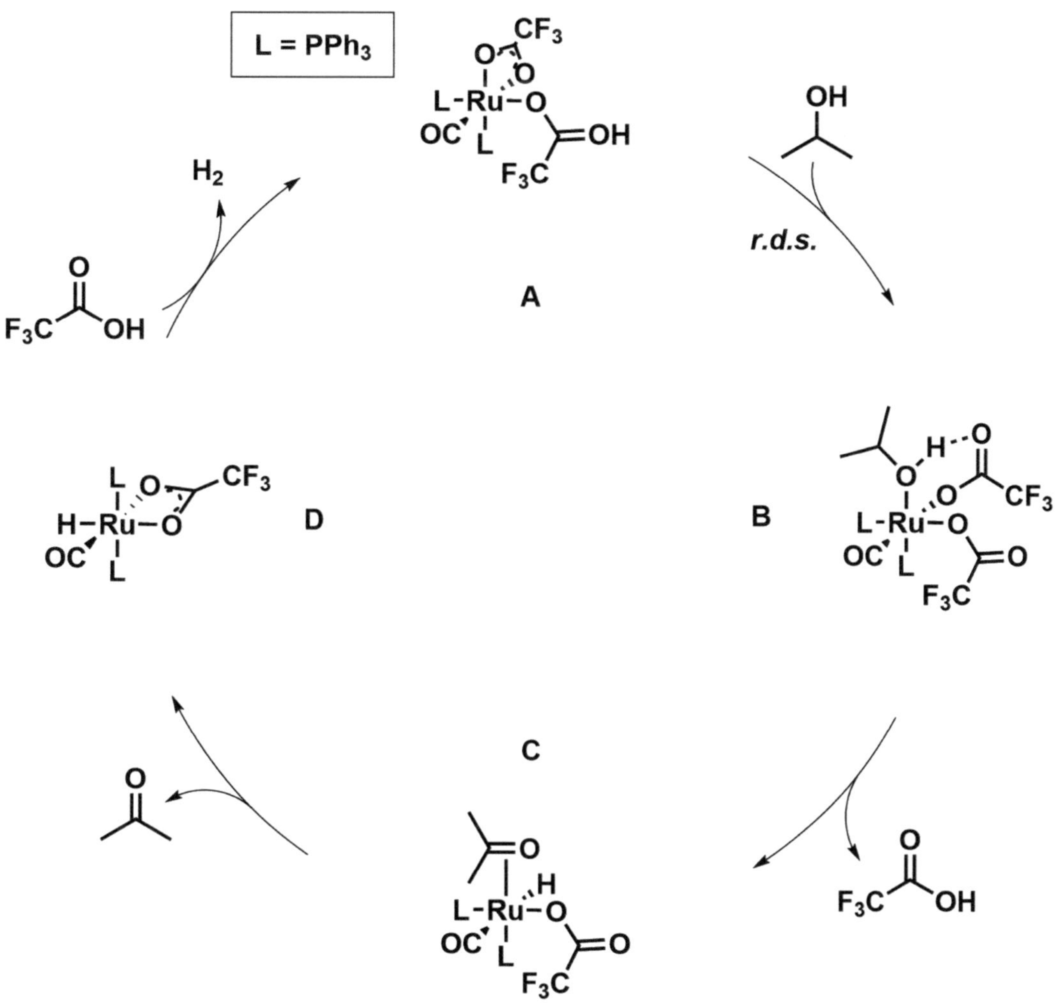

Figure 1. Main proposed elemental catalytic steps of the Robinson isopropanol AAD system. Best result: TOF = 13 h^{-1}.

The steps **B** to **C** and **D** to **A** both involve trifluoroacetic acid as a crucial player. In the former, the acid is extruded, whereas in the latter, it is used as hydride protonation agent and for the re-coordination of trifluoroacetate. Hence, the two steps prefer a low and high acid concentration, respectively, explaining an optimal situation of the 12 equivalents to the Ru complex with respect to obtaining the highest TOF. Moreover, at high acid concentration, the alcohol coordination step (**A** to **B**) appeared to be the rate-determining step.

Contrary, the Cole-Hamilton approach utilises an alkaline-based mechanism, as outlined in **Figure 2**. Commencing with complex **A**, that is, the result of N_2 dissociation from $RuH_2(N_2)(PPh_3)_3$, isopropylate coordinates leading to anionic intermediate **B**. A β-hydride elimination and acetone dissociation then form the trihydride complex **C**, which is sufficiently basic to deprotonate isopropanol to yield the dihydride dihydrogen species **D** and regenerate isopropylate. Finally, loss of H_2 closes the catalytic cycle. This step was found to be the rate-determining, which was corroborated by an incremental effect on the TOF by applying a 500 W tungsten halogen light source.

In 2005 [6] and 2007 [7], Beller developed isopropanol AAD systems based on mixtures of a ligand and either a Ru(II) or Ru(III) precursor in refluxing isopropanol containing 0.8 M NaO*i*Pr. Thus, using two equivalents of 2-di-*t*butyl-phosphinyl-1-phenyl-1*H*-pyrrole to 315 ppm of $RuCl_3$ resulted in a TOF of 155 h^{-1} after 2 h (TOF_{2h}) [6], and five equivalents of tetramethylethylenediamine (TMEDA) to 2.0 ppm of $[RuCl_2(p\text{-cymene})]_2$ gave a TOF_{2h} of 519 h^{-1} [7].

The authors note a clear catalyst activity dependence on the loading of $[RuCl_2(p\text{-cymene})]_2$ [7]. Hence, with 8 and 40 ppm of $[RuCl_2(p\text{-cymene})]_2$ with 0.5 equivalents TMEDA, TOF_{2h}'s of 309 and 161 h^{-1}, respectively, are observed. Therefore, similar activities are obtained when employing 315 ppm ($RuCl_3$/two 2-di-*t*butyl-phosphinyl-1-phenyl-1*H*-pyrrole) and 40 ppm

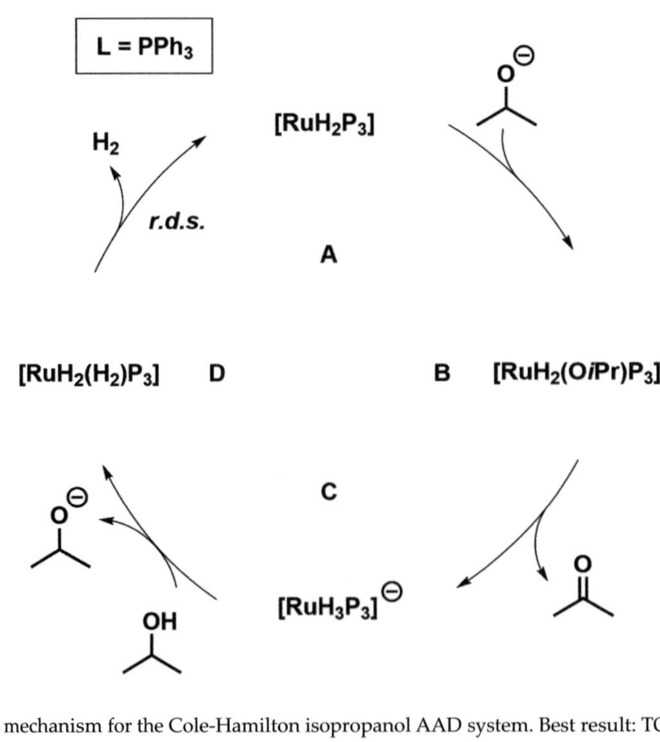

Figure 2. Proposed mechanism for the Cole-Hamilton isopropanol AAD system. Best result: TOF = 330 h^{-1}.

([$RuCl_2$(*p*-cymene)]$_2$/0.5 TMEDA) (155 vs. 161 h^{-1}, respectively). The authors did not carry out activity studies using similar Ru loadings, and it is therefore not possible to make a direct comparison between the two systems. The trend of the catalyst loading effect in the 2007 paper does suggest, nevertheless, that this system would in fact be less active than the 2005 system when using similar Ru loadings.

The fact that lowering the catalyst loading of [$RuCl_2$(*p*-cymene)]$_2$/TMEDA leads to a more active catalytic system might indicate that an associative/dissociative process is involved (*vide infra*) [10]. However, this was not discussed in the paper [7].

Moreover, the 2.0 ppm ([$RuCl_2$(*p*-cymene)]$_2$/five TMEDA) system was allowed to run for 11 days, at which point it was still active and it had reached a turnover number (TON) of 17,215, corresponding to an overall TOF of 64 h^{-1} [7]. No mechanism was proposed.

In 2011, Beller reported on a bis-isopropyl phosphorous substituted phosphorous-nitrogen-phosphorous (PNP^{iPr}, see **Figure 3**) pincer ruthenium catalyst [RuH_2(PNP^{iPr})CO] [8, 9] that, when formed *in situ* from mixing 1:1 [$RuH_2(PPh_3)_3CO$] and PNP^{iPr} ligand, dehydrogenates isopropanol *via* a proposed outer-sphere β-hydride elimination contrary to the until then suggested inner-sphere approaches [9]. The PNP ligand holds an amine unit which deprotonates under somewhat mild conditions leading to an amide-ruthenium bond. Hence, the ligand plays a cooperative role during the catalytic cycle. This setup allowed for conducting the AAD under neutral conditions without using any additives.

Moreover, this led to a drastic increase in TOF with an observed TOF_{max} of 14,145 h^{-1} when employing a 4.0 ppm loading of the catalyst in refluxing isopropanol. This corresponds to a more than 25-fold increase in catalyst turnover frequency compared to the previous state-of-the-art [7]. Interestingly, adding merely 1.3 equivalent of Na*i*OPr in fact led to an approximately 10% decrease in TOF. Moreover, screening results suggested that the CO was vital for obtaining any AAD activity and that exchanging one of the hydrides with a chloride rendered the addition of 1.3 equivalent of Na*i*OPr necessary. The latter suggests that the role of the base in this case is to eliminate off the chloride, thus generating complex **A** in **Figure 3**.

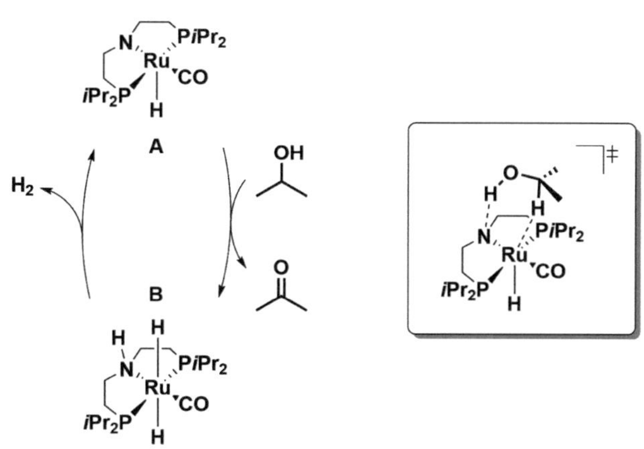

Figure 3. Proposed mechanism for the Beller isopropanol AAD system. Best results: TOF = 14,145 h^{-1}. TON > 40,000. Stable for more than 12 h.

When increasing the catalyst loading eight times from 4.0 to 32 ppm, a fourfold decrease in TOF_{2h} was observed (8342–2048 h^{-1}). This was not addressed further, but it might be speculated whether a detrimental association of two catalyst molecules is more likely at a higher concentration (*vide infra* for more discussions on this topic) [10].

A TON of more than 40,000 was achieved after merely 12 h, compared to the 17,215 after 11 days in the previous AAD report by Beller [7]. Notably, the system was noted to be still highly active after the 12 h. In this regard, it might be speculated whether the chelating pincer ligand renders the catalytic complex particularly robust and thus enables a prolonged lifetime of the system.

Figure 3 shows the proposed mechanism. Because no additives or protonation/deprotonation steps are involved during the catalytic cycle, the entire cycle is composed of two elemental steps, β-hydride elimination of the isopropanol and dehydrogenation of the catalyst. This fact might contribute to the markedly enhanced catalytic activity.

As mentioned, the two steps involve the outer-sphere β-hydride elimination of isopropanol by complex **A** leading to acetone and complex **B** followed by H_2 formation and extrusion, thus regenerating **A**. According to the suggested mechanism, the ligand nitrogen atom plays a fundamental role in both steps. Hence, during the first step, the amide functionality coordinates to the alcohol proton thereby enhancing the carbon-based hydride abstraction. Similarly, the amine serves as proton transferring source to the alkaline hydride leading to the H_2 production.

Overall, the kinetics and longevity of the catalytic systems for isopropanol AAD seem to be highly influenced by the catalytic mechanism and of the necessity of an additive involved in it. Moreover, it is clearly feasible to effectively both dehydrogenate isopropanol and subsequently extrude H_2 without any additive-mediated catalyst activation. Thus, devising a system that employs as simple a mechanism as possible and that are in the absence of catalytic sinks might be important facets to strive for designing new AAD catalysts in the future.

It remains to be disclosed whether an inner- or outer-sphere β-hydride elimination is the key to reaching the superior catalyst activity demonstrated by Beller. Thus, more investigations on this topic would be interesting.

3. Primary alcohols

3.1. Ethanol

In 1987, Cole-Hamilton demonstrated that EtOH can be dehydrogenated with a TOF of 96 h^{-1} by 1×10^{-3} M (61 ppm) $[Rh(bipy)_2]Cl$ in EtOH containing 5% v/v and 1.0 M NaOH at 120°C [11]. The same group improved on this in 1988 [5] and 1989 [12] with a TOF of 210 h^{-1} by use of 3.48×10^{-4} M (20 ppm) $RuH_2(N_2)(PPh_3)_3$, 1 M NaOH, and an intense light source at 150°C. Mechanistic considerations similar to those described for isopropanol (**Figure 2**) were discussed.

In 2012, Beller demonstrated that the same catalytic system that showed superiority with respect to isopropanol AAD (**Figure 3**) also provide interesting results with ethanol [9] Hence, a TOF_{2h} of 1483 h^{-1} could be achieved when using 3.1 ppm 1:1 $[RuH_2(PPh_3)_3CO]$ and PNP^{iPr} ligand in refluxing neutral ethanol. Both acetaldehyde and ethyl acetate were observed as products in substantial amounts. The former product might explain the decrease in activity (TOF_{6h} = 690 h^{-1}) due to a reversible dehydrogenation/hydrogenation process. Moreover, a similar mechanistic rationale as depicted in **Figure 3** was provided.

A constant catalyst activity towards full conversion of ethanol to ethyl acetate can be achieved by adding a minute amount of Na*i*OPr [13] Hence, when using 50 ppm of the commercially available Ru-MACHO ([RuHCl(PNP^{Ph})CO]) in refluxing ethanol for 46 h in the presence of 0.6 mol% NaOEt, a 77% yield of ethyl acetate (TON = 15,400) was obtained. This could be increased slightly to 81% when using 500 ppm catalyst loading and 1.3 mol% NaOEt. Interestingly, a yield of 70% was obtained when conducting the reaction at merely 70°C.

Studies into the effect of additive composition were undertaken. This provided two main results. First, NaOEt was superior to KOEt, NaOH, K_2CO_3 and Cs_2CO_3. Second, an optimal NaOEt loading with respect to maximising the TOF was observed.

Moreover, the 1:1 [$RuH_2(PPh_3)_3CO$]/PNP^{iPr} ligand combination showed similar activity to Ru-MACHO with TOF's of the former of 1107 and the latter of 1134 h^{-1} when employing 25 ppm catalyst loading and 1.3 mol% NaOEt.

Notably, with Ru-MACHO, the conversion rate is practically constant until 90% of the ethanol is used up, at which point the NaOEt is precipitating out of the reaction. This resulted in a TOF_{2h} of 934 h^{-1} and TOF_{10h} of 730 h^{-1} when using 50 ppm catalyst loading. Hence, it was concluded that the reverse hydrogenation process of ethyl acetate was occurring at a negligible level.

Again, a similar mechanism to the one depicted in **Figure 3** was suggested. However, as shown in **Figure 4**, this now involved the dehydrogenation of two different species. Hence, initially ethanol is dehydrogenated into acetaldehyde, which then reacts with either an ethanol or an ethoxide to generate a hemiacetal or anionic hemiacetal intermediate. This compound then undergoes the second dehydrogenation step, leading to another H_2 molecule and the ethyl acetate product.

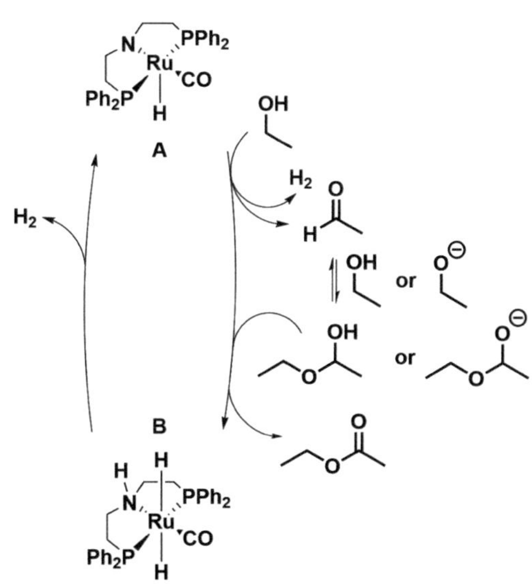

Figure 4. Proposed mechanism for the Beller ethanol AAD to ethyl acetate system. Best result: TOF = 1137 h^{-1}. TON = 15,400. Yield = 81%. Stable for more than 46 h.

In 2012, Gusev reported that the NNP ruthenium species **2**, depicted in **Figure 5**, is superior to Ru-MACHO [10] The Gusev setup employs a slightly modified setup compared to the Beller setup (i.e., less NaOEt, and shorter reaction times than in the Beller setup), and a direct comparison of the results is therefore not feasible. Nevertheless, Gusev also tested the Ru-MACHO catalyst with his system, allowing for direct comparisons of the catalysts.

Thus, in the Gusev setup, 50 ppm Ru-MACHO in refluxing ethanol containing 1 mol% NaOEt led to 42% conversion after 40 h. An impressive 85% conversion was observed under the same conditions with complex **2**, clearly demonstrating the superiority of this catalyst. Moreover, a TOF_{24h} was 567 h^{-1}. Furthermore, 50 ppm **2** was able to reach 83% conversion of simple anhydrous commercial ethanol to ethyl acetate in the presence of air. Hence, air or water seems to have no detrimental influence on catalyst activity. In this regard, it should be noted that all other the reactions performed by Gusev and Beller are done under rigorously inert conditions.

A rather extensive catalyst screening was performed with ruthenium and osmium as transition metals, which **Figure 5** shows a selection of. A quick survey reveals that Ru-MACHO, **1**, and **3** perform similar levels despite **3** being significantly different than the two former complexes. Moreover, changing the *P*-substituents from phenyl to isopropyl seems to have no influence of the conversion, which is in line with the Beller findings. On the other hand, exchanging the CO in **3** to a PPh_3 in **2** drastically increases the conversion, which is contrary to the Beller observations that concluded the CO to be crucial for achieving activity. Moreover, when employing the trihydride osmium(IV) species **4**, practically no conversion is observed.

The influence of the metal can be directly compared in complexes **1** and **5** and in **3** and **6**. Thus, the PNP ruthenium-based **1** shows a slightly superior activity than the osmium-based congener **5** (47% conversion versus 35% after 24 h), whereas the PNN ruthenium-based **3** is noticeably more active than the osmium-based **6** (42 versus 13%). Thus, the PNN ligand seems

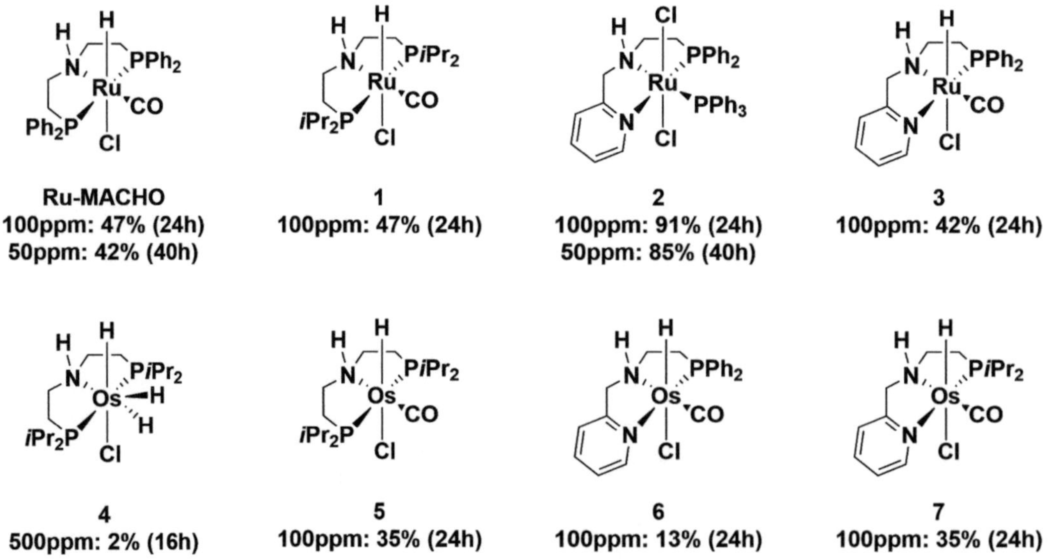

Figure 5. Selection of the complexes tested for ethanol AAD to ethyl acetate by Gusev. Best result: 85% conversion with 50 ppm after 40 h.

http://dx.doi.org/10.5772/intechopen.70654

considerably more susceptible to the choice of metal than the PNP ligand, suggesting that two, or more, different mechanisms are operating.

Finally, the low conversion obtained with complex **6** could be improved on by simply exchanging the *P*-phenyl substituents with isopropyl (13% with **6** versus 35% with **7**).

A mechanism akin to the depiction in **Figure 4** was suggested. Furthermore, H_2 extrusion from the hydrogenated form of **2** was proposed. Moreover, as Beller observed for isopropanol AAD [9], it was noted that decreasing the catalyst loading had a beneficial effect on the TOF. As such, the TOF_{24h} was 375 h^{-1} with 100 ppm and 567 h^{-1} with 50 ppm. It was in this respect suggested that an associative/dissociative process was involved. Varying the loading of the osmium dimer in **Figure 6** further corroborates such a process. Thus, when reducing the catalyst loading from 500 to 100 ppm, the TOF likewise increased approximately five-fold (56–275 h^{-1}). This has the striking consequence that after 24-h reaction time, the 100 ppm loaded mixture afford 66% conversion, whereas the 500 ppm only provide 45%.

In 2014, Beller demonstrated that bioethanol can be effectively converted to acetate by AAD [14]. The complex [RuHCl(PNP^{iPr})CO] provide the best catalyst turnover, and a TOF_{1h} of 1770 h^{-1} is observed when employing 25 ppm catalyst loading in refluxing wet bioethanol containing 8 M NaOH. This result is similar to that found when employing dry ethanol [9] (1770 versus 1483 h^{-1}) albeit at severely harsher conditions. The highly alkaline media was necessary to maintain the product in a deprotonated state, presumably to avoid catalyst deactivation by coordination of acetic acid to the catalyst. Moreover, a 70% yield was obtained within 20 h when using a 1:1 EtOH/H_2O mixture. In addition, a long-term reaction with 10 ppm catalyst loading reached a TON 80,000 after 98 h.

Overall, the results with ethanol clearly demonstrate that primary alcohols are notoriously more difficult to achieve high TOF with than with secondary congeners. Thus, when comparing state-of-the-art turnover frequencies of ethanol AAD (1770 h^{-1}) [14] with that for isopropanol (14,145 h^{-1}) [9], there is an order of magnitude difference in favour of the latter.

Moreover, there is still a lack of studies into the mechanism of the various discrete catalytic steps. Shedding light on these would provide a deeper insight into the kinetic features and parameters of primary alcohol AAD by homogeneous catalysis.

3.2. Methanol

In 1987, Cole-Hamilton demonstrated that MeOH can be dehydrogenated with a TOF of 7 h^{-1} by 1×10^{-3} M (43 ppm) [Rh(bipy)$_2$]Cl in MeOH containing 5% (v/v) H_2O and with 1.0 M NaOH at 120°C [11]. This was the year later improved to 37.3 h^{-1} by the same group by use of 1–5 ×

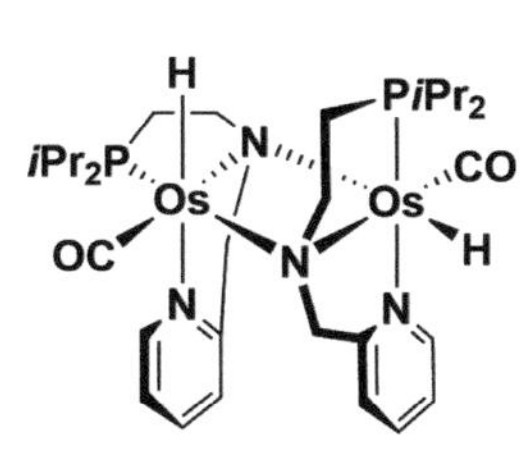

Figure 6. PNN osmium dimer by Gusev. 500 ppm: 45% conversion (24 h), 100 ppm: 66% conversion (24 h).

10^{-4} M (4–20 ppm) $RuH_2(N_2)(PPh_3)_3$, 1 M NaOH, and an intense light source at 150°C [5]. A mechanism as depicted in **Figure 2** was proposed.

In 2013, Beller disclosed a procedure for homogeneously catalysed aqueous-phase reforming type conversion of $MeOH/H_2O$ mixtures to $3H_2$ and CO_2 (or other C_1 residuals, such as carbonate, see **Figure 7**) [15]. Using 1.6 ppm of [RuHCl(PNP^{iPr})CO] in MeOH with 8.0 M KOH at 95.0°C afforded a TOF_{1h} of 4719 h^{-1}. Furthermore, using 19 ppm of [RuHCl(PNP^{Ph})CO] with respect to MeOH in a 9:1 (v/v) $MeOH/H_2O$ mixture afforded a TOF_{1h} of 63 h^{-1} at 65°C. As a note, the TOF was counted in such way that a complete reaction of $MeOH/H_2O$ mixtures to CO_2 and $3H_2$ sums as three turnovers. This was done because all three reactions depicted in **Figure 7** occurs simultaneously, rendering any quantitative kinetic discrimination between them unpractical.

The system turned out to be very robust, with a TON over 350,000 and reaction time exceeding 23 days when using 1 ppm catalyst loading with respect to MeOH of [RuHCl(PNP^{iPr})CO] in a refluxing 9:1 (v/v) $MeOH/H_2O$ solution containing 8.0 M KOH. Moreover, after the 23 days a 27% yield of full MeOH reforming was achieved (based on H_2 evolution and yield based on H_2O as the limiting factor. The yield is 12% with respect to MeOH). When using 150 ppm, a CO_2-based yield of 43% was reached within 24 h (yield based on H_2O as the limiting factor. The yield is 19% with respect to MeOH).

It was also demonstrated that a continuous production of a 3:1 H_2/CO_2 gas mixture, and hence full MeOH reforming, can be achieved by employing 250 ppm catalyst loading with respect to MeOH of the [RuHCl(PNP^{Ph})CO] in a refluxing 4:1 (v/v) $MeOH/H_2O$ solution containing 0.1 M NaOH. After an initiation time of approximately 5–6 h, the expected 3:1 ratio of H_2 and CO_2 was observed in the gas mixture. In addition, the pH dropped from 13 to approximately 10 during the first 4 h. It was suggested that during this initiation time, the hydroxide was reacting with formic acid and CO_2 leading to an eventual equilibrium between hydroxide/(bi)carbonate/formate as the C_1 residuals.

The catalyst activity was depending on a range of factors. Besides the reaction temperature the pH, base additive, and catalyst loading all influenced the activity. As such, a higher pH and lower catalyst loading promoted an increased turnover frequency. The latter is in agreement

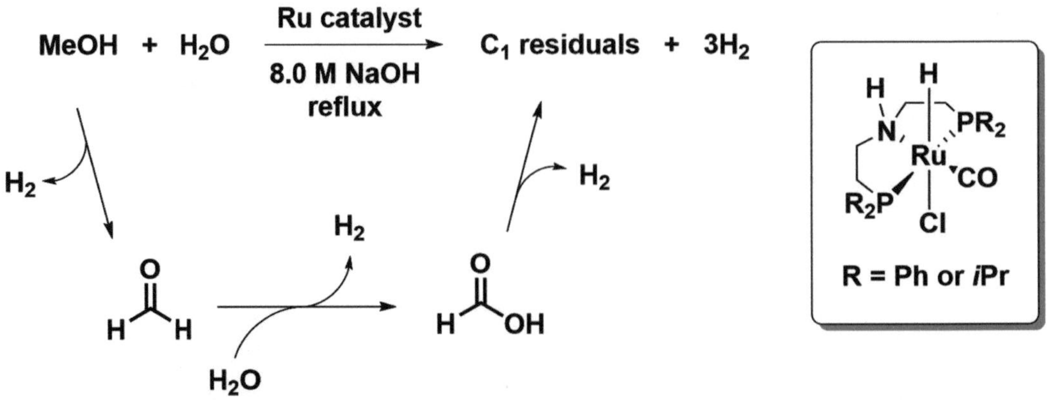

Figure 7. Aqueous MeOH AAD to $3H_2$ and C_1 residuals by Beller. Best results: TOF = 4719 h^{-1}. TON > 350,000. Yield = 43%. Stable for more than 23 days.

with previous observations by, e.g., Beller [9] and Gusev [10]. Moreover, the requirement of a high pH to induce high catalyst activity might reflect the tendency of the catalyst to reside in a range of resting states, particularly with a coordinating formic acid. In order to re-activate the catalyst, a base can eliminate off, e.g., the formic acid from the resting catalyst.

Catalyst TOF dependency on pH is likely also a major reason for the influence of base additive. However, Bernskoetter, Hazari, and Holthausen demonstrates in a later publication that the cationic counter ion might very well play a crucial role as well (*vide infra*) [16]. Even though they employ modified reaction conditions, the same effect of the cation might also be in play in the Beller setup.

An in-depth study revealed several aspects of the mechanism(s) [17]. An Arrhenius plot revealed the temperature-activity dependency, and with [RuHCl(PNP^{iPr})CO] an activation energy of E_a = 82.4 kJ/mol and A = 1.2 × 10^6 mol/s were found. Furthermore, a kinetic isotope effect (KIE) of 7.07 was observed, strongly suggesting proton involvement in the rate-determining step. However, because three reactions are concomitantly taking place, any further conclusions on the mechanism are difficult. Moreover, at certain catalyst loadings the gas evolution initially follows pseudo zero kinetics. The same incremental effect on TOF upon decreasing the catalyst loading was observed as well, providing a reaction order with respect to the catalyst of less than 1. Finally, computational studies were employed to shed further light on the mechanism(s).

This led to a revised suggested mechanism. Overall, the ruthenium-amido functionality still plays a key role, but an inner-sphere mechanism for the β-hydride elimination involving, e.g., a methoxide for the MeOH dehydrogenation step was discussed and proposed. In addition, the dehydrogenation step is assisted by a MeOH molecule, akin to previously described by Schneider [18]. Hence, in the latter step, a transient protonation of one of the hydrides may be involved.

Interestingly, the *N*-methylated congener to the isopropyl *P*-substituted catalyst was tested as well. Considering the key role of the amine/amido unit of the so far proposed mechanisms, a drastic drop in TOF was expected. However, surprisingly a mere drop of 2.4 times in catalyst activity was observed. However, a KIE of 1.76 suggests a change in mechanism. Furthermore, a bell-shaped activity dependency on KOH concentration with a maximum activity at 4.0 M KOH was observed. In fact, at this base concentration, the *N*-methylated catalyst is almost twice as active than the original one at 60°C (approximately 100 versus approximately 50 h^{-1}) and approximately 50% more active at 90°C (approximately 200 versus approximately 125 h^{-1}).

These results all clearly point towards a change in mechanism upon methylating the ligand nitrogen atom. Moreover, computational studies suggested that a higher stability towards hydride protonation was responsible for the bell-shaped activity-base concentration behaviour.

Beller later showed that by mixing Ru-MACHO-BH (chloride of Ru-MACHO exchange with a borohydride) with [Ru(H)$_2$(dppe)$_2$], MeOH reforming can be achieved under neutral conditions [19]. Thus, mixing 22.5 ppm with respect to MeOH of each of the catalysts in a 9:1:4 (v/v) mixture of MeOH/H_2O/triglyme at 93.5°C applied temperature afforded a TOF_{1h} of 87 h^{-1}. A long-term experiment afforded a 26% yield to H_2O, corresponding to a TON > 4200.

Even though this value is considerably lower than for the system containing base, it still proofs the principle of base-free MeOH reforming. Interestingly, the combination of the two catalysts provided a system significantly more active than the sum of the two catalysts individual performance.

In 2013, Grützmacher devised another catalytic system for methanol reforming by homogeneously ruthenium catalysed AAD under neutral conditions [20]. Conducting the MeOH reforming using 500 ppm of [K(dme)$_2$][Ru(H)(trop$_2$dad)] (**A** in **Figure 8**) in refluxing THF containing a 1:1.3 mixture of MeOH/H_2O (90°C applied temperature) afforded 90% conversion after 10 hours, corresponding to an overall TOF of 54 h^{-1} and TON of 540. Moreover, the yield was 84% yield.

The proposed mechanism using [K(dme)$_2$][Ru(H)(trop$_2$dad)] is markedly different from that using [RuHCl(PNP^{iPr})CO] as catalyst. As shown in **Figure 8**, the ruthenium is redox active during the catalytic cycle. Hence, commencing with complex **A** ([Ru(H)(trop$_2$dad)]$^-$) containing a Ru^{II} center, a water assisted hydride protonation and subsequent dehydrogenation to species **B** with unspecified oxidation state is occurring. MeOH then adds to the Ru-N bond affording complex **C**, which has a Ru^{II} center. A β-hydride elimination then leads to the extrusion of formaldehyde, which reacts with water to give methanediol. Furthermore, the metal center is reduced to Ru^0 and the singly protonated 1,2-enediamide moiety in **C** becomes further protonated to yield the amino imine complex **D**. The methanediol is then dehydrogenated by **D** to yield formic acid and the imine moiety in **D** is reduced to afford diamine complex **E**, which upon de-coordination of the formic acid leads to species **F**. Hence, at this stage, complex **B** has taken up two equivalents of H_2. Finally, a base-assisted dehydrogenation of the ligand framework and consequently oxidation of the ruthenium converts **F** back to **A**, thereby closing the catalytic cycle.

Computational studies indicate that the conversion of methanediol to formic is faster than the conversion of MeOH to methanediol, explaining the absence of formaldehyde during the reaction. Moreover, it was demonstrated that complex **A** efficiently catalyses the dehydrogenation of formic acid to H_2 and CO_2. Thus, employing a 100 ppm catalyst loading in a 1 M formic acid solution in dioxane at 90°C provided an initial TOF of 24,000 h^{-1}.

In 2014, Milstein also disclosed a catalytic system for MeOH reforming by AAD [21]. A catalyst loading of 250 ppm (with respect to MeOH) of the PNN ruthenium complex shown in **Figure 9** in a 5.55:1 mixture of MeOH/H_2O in toluene at 100–105°C (115°C applied temperature) in the presence of two equivalents of KOH with respect to MeOH was employed. This led to a H_2-based yield of 77% after 9 days. Interestingly, the organic layer of the reaction could be isolated and reused for another round of MeOH reforming. Doing so twice led to an overall TON of approximately 29,000 with a yield after the third round of 80%, again after 9 days. Hence, the system seems feasible for reusing, which is an important factor for developing applicable MeOH reforming systems.

The decomposition of formic acid to H_2 and CO_2 was also studied. When employing 900 ppm of the catalyst, 1.2 equivalents of KO*t*Bu with respect to catalyst, and pure formic acid in a 1:1 (v/v) THF/H_2O mixture a mere 25% conversion was observed after 24 h. This was improved to >99% upon exchanging the KO*t*Bu with two equivalents of KOH. Interestingly, when no H_2O was present, a reaction containing two equivalents of Et_3N as base leads to 98% conversion after 24 h at room temperature. It was thus concluded that formic acid decomposition is markedly more facile than its formation from methanol. Hence, the first two steps of the MeOH reforming were suggested to be rate-determining. Moreover, mechanistic studies suggest that one of the ligand methylene hydrogens takes part of the catalytic cycle *via* dearomatization of the central pyridine unit, such as Milstein has previously demonstrated with other AAD systems [22].

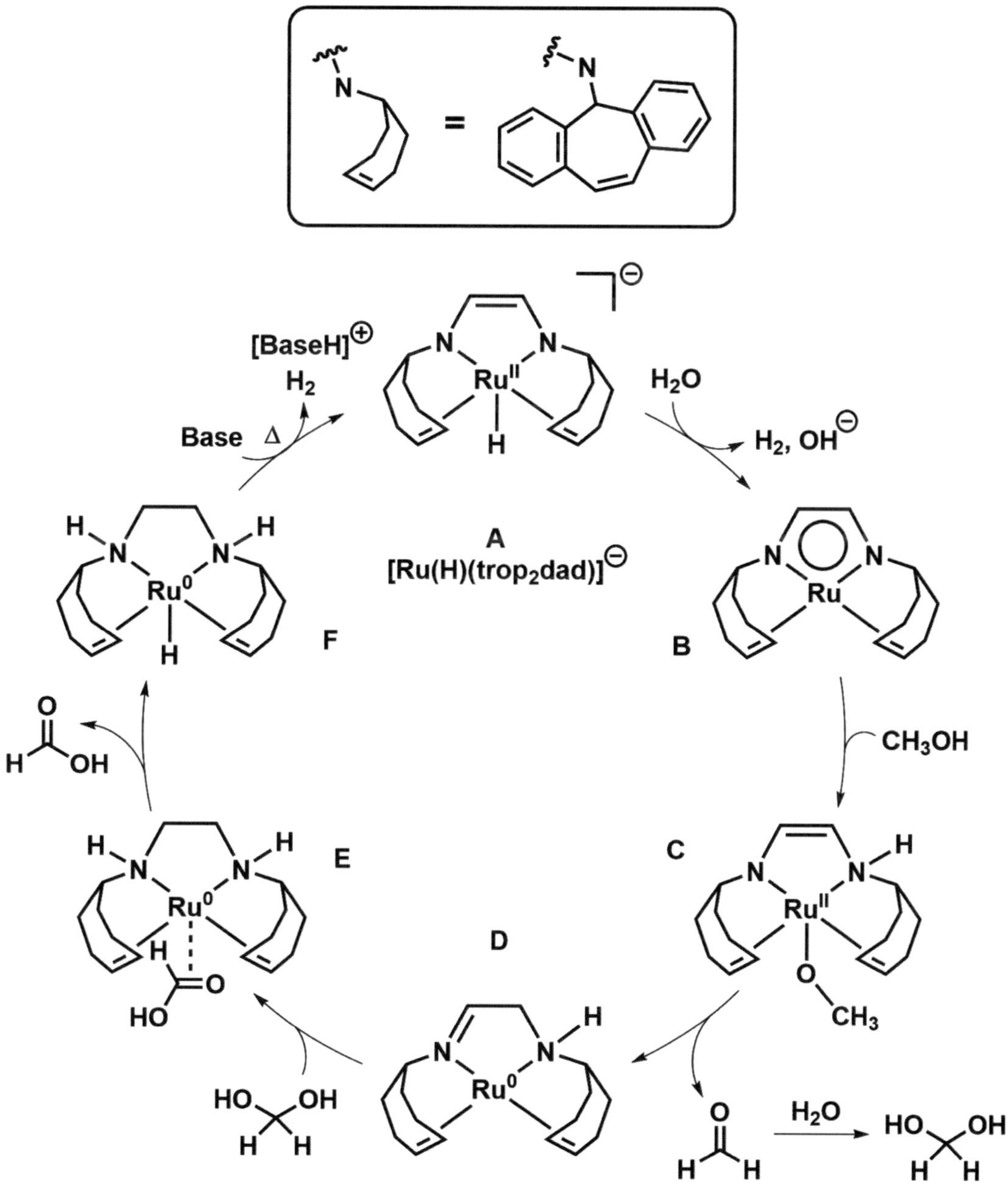

Figure 8. Proposed catalytic cycle for MeOH reforming by Grützmacher using MeOH as substrate. Best results: TOF = 54 h^{-1}. TON = 540 h^{-1}. Conversion = 90%. Yield = 84%.

The Reek group showed in 2016 that [Ru(salbinapht)(CO)(P*i*Pr$_3$)] (**Figure 10**) is also capable of dehydrogenating MeOH [23]. A TOF value of 55 h^{-1} could be reached by a catalyst loading of 24 ppm with respect to MeOH in a 25% dioxane/75% 9:1 (v/v) MeOH/H_2O mixture containing 8 M KOH and conducting the reaction at 82°C. Lowering the base concentration to 6 M and

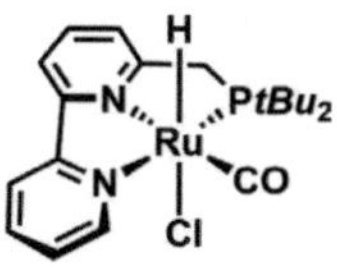

Figure 9. Milstein PNN ruthenium catalyst for MeOH reforming. Best results: TOF = 45 h^{-1}. TON = 29,000. Yield = 82%. Reusable system.

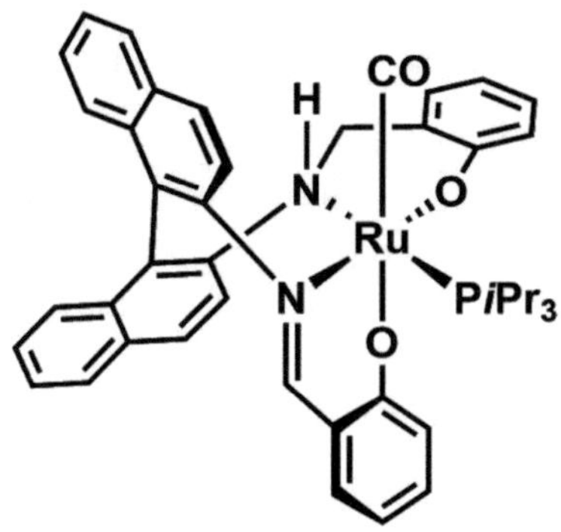

Figure 10. Reek catalyst for MeOH reforming. Best result: TOF = 55 h^{-1}.

4 M afforded TOF's of 37 and 29 h^{-1}, respectively. This was partially explained by the decrease in reaction temperature as a result of the lower base concentration (79°C with the 6 M and 76°C with the 4 M).

It was also revealed that $[Ru(H)_2\,(PiPr_3)_2(CO)_2]$ catalyses the reaction with practically the same efficiency (TOF = 50 h^{-1}), leaving some speculations as to whether this is the real catalyst.

Moreover, mechanistic investigations suggest that the CO moiety of [Ru(salbinapht)(CO)($PiPr_3$)] plays an active role during the catalytic cycle by reacting with hydroxide and thus forming formic acid and H_2. This might open for the possibility that a similar mechanism might (partially) take place with the Beller and Milstein systems.

Other metal complexes have also been shown to conduct MeOH reforming by AAD, specifically iron [16, 24], manganese [25], and iridium [26–28]. Several of them are comprised of PNP^{iPr} pincer ligand complexes, such as the iron-based compounds shown in **Figure 11**. Beller initially showed that borohydride coordinated species afforded a TOF_{2h} of 644 h^{-1} when using a 4.5 ppm catalyst loading in a 9:1 (v/v) $MeOH/H_2O$ mixture containing 8.0 M KOH and stirring at 91°C [24]. A long-term experiment allowed for a TON_{46h} of 9834.

When comparing this result using the iron-based catalyst with the TOF values obtained when using the ruthenium-based congener, the latter is superior with respect to catalyst activity and longevity. Nevertheless, showcasing the feasibility of conducting MeOH reforming using a non-noble metal catalyst is an important step towards applicability, which these findings therefore represent.

Bernskoetter, Hazari, and Holthausen improved the iron-based MeOH reforming by exchanging the borohydride with a formate, dissolving minute amounts of MeOH and H_2O in EtOAc,

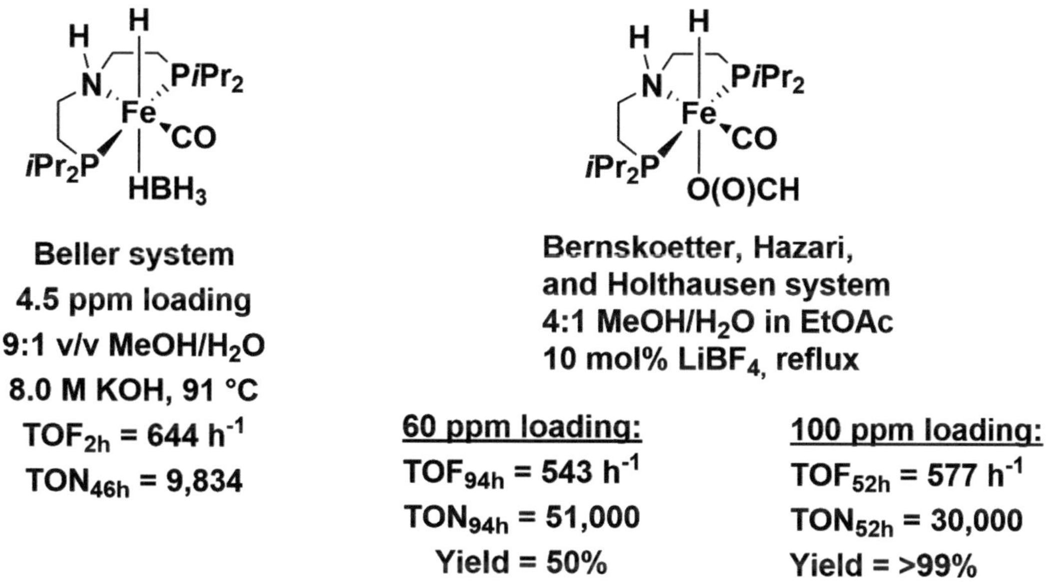

Figure 11. MeOH reforming using PNP iron catalysts.

and adding 10 mol% $LiBF_4$ as additive rather than the 8.0 M KOH [16]. By doing so, an impressive TON of 51,000 could be reached after 94 h by using 60 ppm catalyst loading, which corresponds to 50% yield. A >99% yield could be obtained by increasing the catalyst loading to 100 ppm (TON_{52h} = 30,000).

Based on a number of in-depth studies, including computational insights, the authors suggest that the Lewis acidic Li^+ promotes de-coordination of formate and subsequent its dehydrogenation as well, thereby facilitating a faster formic acid decomposition to H_2 and CO_2. Furthermore, they conclude that under neutral conditions, the formation of the methanediol intermediate is mediated by the catalyst as well.

Interestingly, when using dry MeOH in EtOAc, methyl formate can be formed in >99% yield when otherwise similar conditions as the latter mentioned above, thus corresponding to a TON of >19,999.

Recently, Beller used 0.05 mM concentration a [MnBr(PNP^{iPr})$(CO)_2$] complex and additional 10 equivalents of the PNP^{iPr} ligand in a 1:1 (v/v) mixture of 9:1 MeOH/H_2O/triglyme at a applied temperature of 92°C to achieve a TON of more than 20,000 after more than a month. Thus, the system is highly stable albeit with a low turnover frequency. Moreover, the catalytic cycle was suggested to follow the general mechanism depicted in **Figures 3** and **4**.

Fujita and Yamaguchi demonstrated in 2015 that the iridium complex shown in the catalytic cycle in **Figure 12** facilitates MeOH reforming. The best conditions were found to be refluxing a 1:4 MeOH/H_2O mixture containing 0.50 mol% NaOH. Employing a 5000 ppm catalyst loading afforded a 84% yield after 20 h, corresponding to a TON of 5040 and an overall TOF of 252 h^{-1}.

It was furthermore demonstrated that a system containing 1000 ppm is capable of continuously dehydrogenate MeOH. Hence, when a mixture of MeOH, H_2O, and NaOH was continuously

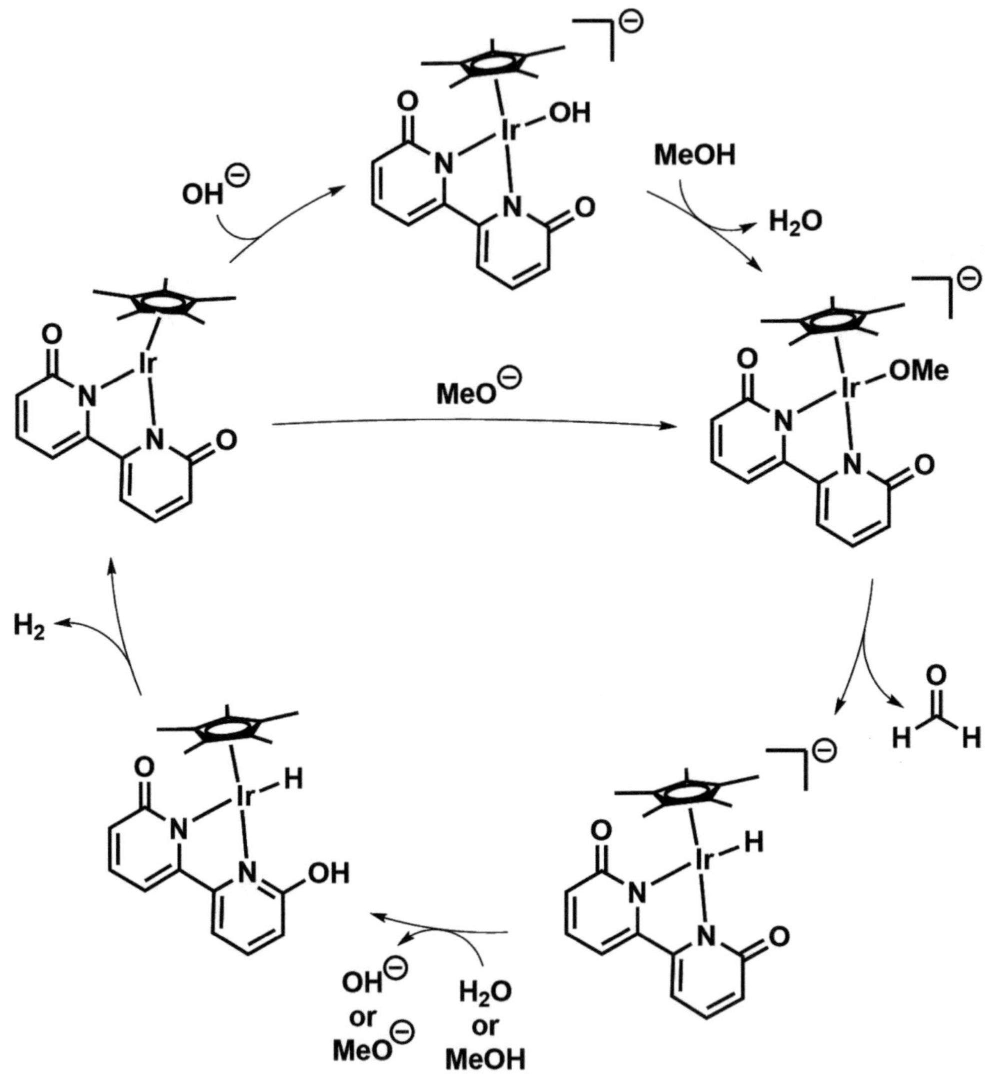

Figure 12. Proposed catalytic cycle for the Yamaguchi system for MeOH reforming.

added dropwise to the solution, a TON of 10,510 was reached after 150 h. This is an important achievement since devising a MeOH reforming system capable of continuously converting MeOH is of application-wise interest.

The catalytic cycle was suggested to follow the mechanism shown in **Figure 12**. Initially, the starting organometallic complex has its hydroxide replaced by a MeOH molecule, which then undergo β-hydride elimination yielding a formaldehyde molecule, that is extruded, and the anionic hydride complex. It is not discussed whether an inner- or outer-sphere β-hydride elimination takes place.

The anionic hydride complex is then prone to protonation at one of the pyridonate moieties, which affords the neutral hydride complex that then undergoes dehydrogenation resulting

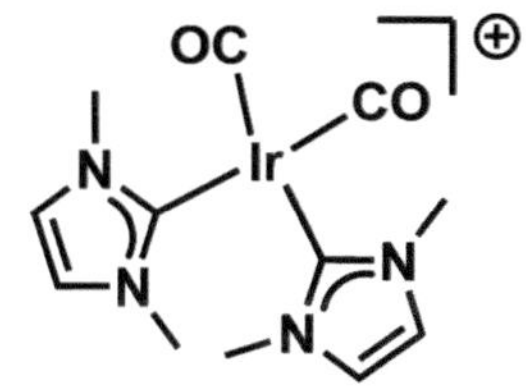

Figure 13. Crabtree catalyst for MeOH dehydrogenation. Best results: TOF_{40h} = 200 h^{-1}. TON_{40h} = 8000.

in the unsaturated tri-coordinated species. Finally, either a hydroxide re-makes the starting catalytic complex, or a methoxide brings the complex straight to the methoxo complex.

The same year, Crabtree demonstrated that the biscarbene iridium complex shown in **Figure 13** is also able to dehydrogenate MeOH [27]. Thus, after 40 h 10 ppm catalyst loading had converted refluxing dry MeOH containing 6.7 M KOH with a TON of 8000, corresponding to an overall TOF of 200 h^{-1}. Moreover, the reaction produced H_2 in a practically linear fashion the first 20 h. Furthermore, the system was found to work in the presence of air, and thus no inert atmosphere was necessary.

Recently, Beller employed a PNP^{iPr} pincer ligated iridium complex for MeOH reforming [28]. At low base concentrations (0.1 M KOH), 19 ppm of $Ir(H)_3(PNP^{iPr})$ in refluxing 9:1 (v/v) MeOH/H_2O (69°C) dehydrogenate MeOH with a TOF_{1h} of 525 h^{-1}. After 16 h, a TON of 1600 was reached after which gas evolution ceased. This could be improved by increasing the base concentration to 8.0 M KOH, in which the system becomes stable for more than 60 h and results in a TON of 1900. Moreover, only a slight drop in catalyst activity was observed during this period.

Interestingly, when increasing the base concentrations, the activity drops. The same is true when lowering the base amount to two equivalents with respect to the catalyst. Hence, a rather small concentration of base is optimal for achieving the best catalytic turnover frequency. Thus, a bell-shaped base-activity curve was observed similar to that for the *N*-methylated PNP^{iPr} ruthenium-based catalyst (*vide supra*) [18].

Employing $Ir(H)_2(PNP^{iPr})CO$ instead of $Ir(H)_3(PNP^{iPr})$ resulted in no gas evolution, showing that the CO unit clearly has a detrimental effect to ability of the iridium-based PNP pincer complex to catalyse the MeOH reforming process. Moreover, a mechanism similar to that depicted in **Figures 3** and **4** were proposed.

Overall, aqueous-phase MeOH reforming by use of homogeneously catalysed AAD methods has witnessed great improvements during the last approximately 5 years. A variety of different metal based complexes has been demonstrated to catalyse the reaction with good activity and longevity. Nevertheless, there is still need for more in-depth studies to reveal the factors important for taking the methodology even further towards applicability.

4. Conclusion

Since the last decade, AAD by homogeneous catalysis has witnessed great improvements. Catalyst activity and stability has mainly been investigated with small molecule transformations, such as methanol, ethanol and isopropanol. This is because with larger alcohol substrates, a more organic synthetic and preparative applicability is the main focus.

Moreover, particularly full methanol AAD corresponding to aqueous-phase methanol reforming currently enjoys much attention due to its possible use in renewable energy storage. Hence, since 2013, several interesting systems have been developed for this specific reaction, and many important insights have been disclosed. This sets the stage for taking this chemistry even further to the next level towards industrial applicability.

However, there is still lacking much research into the more subtle mechanism(s) and factors that are decisive for system efficiency and longevity. Furthermore, to date there has not been developed a reversible system for methanol AAD and CO_2 hydrogenation, which is imperative to achieve in order to reach a level ready for application. In addition, even more active systems are required. At best, these systems should work without any sacrificial additives to afford more economically viable reactions.

Acknowledgements

This work was supported by a research grant (19049) from VILLUM FONDEN.

Author details

Martin Nielsen

Address all correspondence to: marnie@kemi.dtu.dk

Technical University of Denmark, Kongens Lyngby, Denmark

References

[1] Olah GA. Towards oil independence through renewable methanol chemistry. Angewandte Chemie, International Edition. 2013;**52**(1):104-107. DOI: 10.1002/anie.201204995

[2] Charman HBJ. Hydride transfer reactions catalysed by metal complexes. Journal of the Chemical Society B: Physical Organic. 1967. pp. 629-632. DOI: 10.1039/j29670000629

[3] Dobson A, Robinson SD. Catalytic dehydrogenation of primary and secondary alcohols by $Ru(OCOCF_3)_2(CO)(PPh_3)_2$. Journal of Organometallic Chemistry. 1975;**87**(3):C52-C53. DOI: 10.1016/S0022-328X(00)88159-0

[4] Dobson A, Robinson SD. Complexes of the platinum metals. 7. Homogeneous ruthenium and osmium catalysts for the dehydrogenation of primary and secondary alcohols. Inorganic Chemistry. 1977;**16**(1):137-142. DOI: 10.1021/ic50167a029

[5] Morton D, Cole-Hamilton DJ. Molecular hydrogen complexes in catalysis: Highly efficient hydrogen production from alcoholic substrates catalysed by ruthenium complexes. Journal of the Chemical Society, Chemical Communications. 1988;**2**(17):1154-1156. DOI: 10.1039/c39880001154

[6] Junge H, Beller M. Ruthenium-catalyzed generation of hydrogen from iso-propanol. Tetrahedron Letters. 2005;**46**(6):1031-1034. DOI: 10.1016/j.tetlet.2004.11.159

[7] Junge H, Loges B, Beller M. Novel improved ruthenium catalysts for the generation of hydrogen from alcohols. Chemical Communications. 2007;(5):522-524. DOI: 10.1039/B613785G

[8] Bertoli M, Choualeb A, Lough AJ, Moore B, Spasyuk D, Gusev DG. Osmium and ruthenium catalysts for dehydrogenation of alcohols. Organometallics. 2011;**30**(13):3479-3482. DOI: 10.1021/om200437n

[9] Nielsen M, Kammer A, Cozzula D, Junge H, Gladiali S, Beller M. Efficient hydrogen production from alcohols under mild reaction conditions. Angewandte Chemie, International Edition. 2011;**50**(41):9593-9597. DOI: 10.1002/anie.201104722

[10] Spasyuk D, Gusev DG. Acceptorless dehydrogenative coupling of ethanol and hydrogenation of esters and imines. Organometallics. 2012;**31**(15):5239-5242. DOI: 10.1021/om300670r

[11] Morton D, Cole-Hamilton DJ. Rapid thermal hydrogen production from alcohols catalysed by [Rh(2,2′-bipyridyl)$_2$]cl. Journal of the Chemical Society, Chemical Communications. 1987;**1**(4):248-249. DOI: 10.1039/C39870000248

[12] Morton D, Cole-Hamilton DJ, Utuk ID, Paneque-Sosa M, Lopez-Poveda M. Hydrogen production from ethanol catalysed by group 8 metal complexes. Journal of the Chemical Society, Dalton Transactions. 1989;(3):489-495. DOI: 10.1039/dt9890000489. http://pubs.rsc.org/en/content/articlelanding/1989/dt/dt9890000489#!divAbstract

[13] Nielsen M, Junge H, Kammer A, Beller M. Towards a green process for bulk-scale synthesis of ethyl acetate: Efficient acceptorless dehydrogenation of ethanol. Angewandte Chemie, International Edition. 2012;**51**(23):5711-5713. DOI: 10.1002/anie.201200625

[14] Sponholz P, Mellmann D, Cordes C, Alsabeh PG, Li B, Li Y, Nielsen M, Junge H, Dixneuf P, Beller M. Efficient and selective hydrogen generation from bioethanol using ruthenium pincer-type complexes. ChemSusChem. 2014;**7**(9):2419-2422. DOI: 10.1002/cssc.201402426

[15] Nielsen M, Alberico E, Baumann W, Drexler H-J, Junge H, Gladiali S, Beller M. Low-temperature aqueous-phase methanol dehydrogenation to hydrogen and carbon dioxide. Nature. 2013;**495**(7439):85-89. DOI: 10.1038/nature11891

[16] Bielinski EA, Förster M, Zhang Y, Bernskoetter WH, Hazari N, Holthausen MC. Base-free methanol dehydrogenation using a pincer-supported iron compound and Lewis acid co-catalyst. ACS Catalysis. 2015;**5**(4):2404-2415. DOI: 10.1021/acscatal.5b00137

[17] Alberico E, Lennox AJJ, Vogt LK, Jiao H, Baumann W, Drexler H-J, Nielsen M, Spannenberg A, Checinski MP, Junge H, Beller M. Unravelling the mechanism of basic aqueous methanol dehydrogenation catalyzed by Ru-PNP pincer complexes. Journal of the American Chemical Society. 2016;**138**(45):14890-14904. DOI: 10.1021/jacs.6b05692

[18] Friedrich A, Drees M, Schmedt auf der Günne J, Schneider S. Highly stereoselective proton/hydride exchange: Assistance of hydrogen bonding for the heterolytic splitting of H_2. Journal of the American Chemical Society. 2009;**131**(48):17552-17553. DOI: 10.1021/ja908644n

[19] Monney A, Barsch E, Sponholz P, Junge H, Ludwig R, Beller M. Base-free hydrogen generation from methanol using a bi-catalytic system. Chemical Communications. 2014; **50**(6):707-709. DOI: 10.1039/C3CC47306F

[20] Rodríguez-Lugo RE, Trincado M, Vogt M, Tewes F, Santiso-Quinones G, Grützmacher H. A homogeneous transition metal complex for clean hydrogen production from methanol–water mixtures. Nature Chemistry. 2013;**5**(4):342-347. DOI: 10.1038/nchem.1595

[21] Hu P, Diskin-Posner Y, Ben-David Y, Milstein D. Reusable homogeneous catalytic system for hydrogen production from methanol and water. ACS Catalysis. 2014;**4**(8):2649-2652. DOI: 10.1021/cs500937f

[22] Milstein D. Discovery of environmentally benign catalytic reactions of alcohols catalyzed by pyridine-based pincer Ru complexes, based on metal-ligand cooperation. Topics in Catalysis. 2010;**53**(13-14):915-923. DOI: 10.1007/s11244-010-9523-7

[23] Van de Watering F, Lutz M, Dzik W, de Bruin B, Reek JNH. Reactivity of a ruthenium-carbonyl complex in the methanol dehydrogenation reaction. ChemCatChem. 2016;**8**(17):2752-2756. DOI: 10.1002/cctc.201600709

[24] Alberico E, Sponholz P, Cordes C, Nielsen M, Drexler H-J, Baumann W, Junge H, Beller M. Selective hydrogen production from methanol with a defined iron pincer catalyst under mild conditions. Angewandte Chemie, International Edition. 2013;**52**(52):14162-14166. DOI: 10.1002/anie.201307224

[25] Andérez-Fernández M, Vogt LK, Fischer S, Zhou W, Jiao H, Garbe M, Elangovan S, Junge K, Junge H, Ludwig R, Beller M. A stable manganese pincer catalyst for the selective dehydrogenation of methanol. Angewandte Chemie, International Edition. 2017;**56**(2):559-562. DOI: 10.1002/anie.201610182

[26] Fujita K, Kawahara R, Aikawa T, Yamaguchi R. Hydrogen production from a methanol-water solution catalyzed by an anionic iridium complex bearing a functional bipyridonate ligand under weakly basic conditions. Angewandte Chemie, International Edition. 2015;**54**(31):9057-9060. DOI: 10.1002/anie.201502194

[27] Campos J, Sharninghausen LS, Manas MG, Crabtree RH. Methanol dehydrogenation by iridium N-heterocyclic carbene complexes. Inorganic Chemistry. 2015;**54**(11):5079-5084. DOI: 10.1021/ic502521c

[28] Prichatz C, Alberico E, Baumann W, Junge H, Beller M. Iridium-PNP pincer complexes for methanol dehydrogenation at low base concentration. ChemCatChem. 2017;**9**(11):1891-1896. DOI: 10.1002/cctc.201700015

Section 2

Kinetics of Nanomaterials

Chapter 7

Oxidation of Glycerol to Lactic Acid by Gold on Acidified Alumina: A Kinetic and DFT Case Study

Thabang A. Ntho, Pumeza Gqogqa and
James L. Aluha

Additional information is available at the end of the chapter

http://dx.doi.org/10.5772/intechopen.70485

Abstract

The aim of this chapter is to present proposed kinetic and density functional theory (DFT) models for the selective oxidation of glycerol to various hydroxy-acids over an acidified Au/γ-Al_2O_3 catalyst. Glycerol oxidation over gold-based catalysts to value-added chemicals continues to attract attention worldwide. Both the kinetics and theoretical mechanisms of this reaction have been reported in the past. However, some of the reported kinetic data was possibly collected under mass transfer limitations. In this case study we demonstrate that if mass transfer is eliminated, a pseudo zero-order model can be fitted to the experimental data with a high degree of correlation. Furthermore, we propose a plausible mechanism of pyruvaldehyde (PA) isomerisation to lactic acid (LAC) over supported molybdenum Lewis acid sites as investigated with density functional theory (DFT) approach. A proposed DFT model suggested that the rate-limiting step in the isomerisation of PA to LAC, catalysed by a Mo Lewis acid-site, could be the dissociation of a proton from an adsorbed water molecule – the protonation step.

Keywords: gold, catalyst, alumina, support-acidity, glycerol-oxidation, lactic acid, kinetics, DFT

1. Introduction

Lactic acid (LAC), as noted by Wee et al. [1], is one of the most valuable chemicals in industry today and is widely used in the food, cosmetic, pharmaceutical, and chemical industries. In the food industry, for example, it may be used as a preservative, an acidulant, or for flavouring, while in the textile, pharmaceutical and chemical industry it is used as a raw material for the production of lactate ester, propylene glycol, 2,3-pentanedione, propanoic acid, acrylic acid, acetaldehyde and dilactide. In fact, LAC continues to receive increased attention for its potential use as a monomer in the production of biodegradable poly lactic acid. It can be produced by either biotechnological fermentation or chemical synthesis, but the former route is receiving considerable interest due to environmental concerns and the limited nature of petrochemical feedstocks. However, fermentation is inherently a slow process. An alternative route to LAC is by processing a 'renewable' resource such as glycerol by means of heterogeneous catalysis. Haruta [2] has reported that gold with particle diameters below 10 nm are surprisingly active for many reactions, such as CO oxidation and propylene epoxidation.

Many aspects of glycerol oxidation by Au have been studied. Ketchie et al. [3] have looked at the effect of Au particle size, particularly on supports such as carbon [4] and titania [5], while Wang et al. [6] have examined the effect of particle shape, and Villa et al. [7] have investigated the role of stabilizers in gold sols as catalysts in the liquid-phase oxidation of glycerol. In addition, Demirel et al. [8] have probed the promotional effect of Pt on Au/C catalysts, and Royker et al. [9] have investigated the promotional effect of Pt on Au/Al_2O_3 catalysts, while other authors have studied the effect of base in the reaction. For example, Chornaja et al. [10] examined the oxidation of glycerol to glyceric acid using Pd catalysts that worked in alkaline media, and Ketchie et al. [5] reported the promotional effect of hydroxyl ions over Au catalysts, while Carretin et al. [11] have analysed the effect of base as a reaction initiator, proving that for Au/C catalysts, the presence of OH^- was mandatory for any meaningful glycerol oxidation to occur. However, there was seemingly very limited study on the effect of surface acidity for this reaction using alumina supports, for which this work is dedicated.

2. Experimental procedures

2.1. Chemical reagents and materials

Commercial γ-Al_2O_3 support (denoted as Degussa 2010, BET specific surface area of 260 $m^2\ g^{-1}$), ammonium molybdate $(NH_4)_6{\cdot}Mo_7O_{24}{\cdot}4H_2O$ from Associated Chemical Enterprises (ACE), chloroauric acid $HAuCl_4{\cdot}3H_2O$ from Rand Refinery (South Africa), NaOH (98%) and nitric acid (65%) from ACE, and glycerol (99.5%) from Rochelle Chemicals were used.

2.2. Preparation of MoO_3/γ-Al_2O_3 support

A mass of 4.6 g of γ-Al_2O_3 support with a measured BET specific surface area of 245 $m^2\ g^{-1}$ was weighed and placed in a beaker. About 100 ml solution of 0.1 M ammonium molybdate salt was measured, enough to form MoO_3 monolayer coverage at surface concentration of 5 atoms of metal nm^{-2} or 0.2 nm^2 $atom^{-1}$ according to Stobbe-Kreemers et al. [12] and Raubenheimer and Cronje [13]. The pH of the solution was then adjusted to a value below 1 by addition of dilute (0.25 M) HNO_3 acid, with agitation to ensure equal distribution of the acid to a stable pH. The support was then added to the ammonium molybdate solution and left to stand for 8 h, after which, it was filtered and left to oven dry in air at 120°C for 16 h. The dried catalyst precursor was then calcined in air at a flow rate of 300 SCCM at 500°C for 4 h to decompose any residual ammonium and nitrate ions from the support, effectively reducing the molybdate ion to MoO_3, thereby anchoring it to the alumina support. Gold was loaded onto the supports as described elsewhere [14].

2.3. Catalyst testing conditions

Catalyst testing for glycerol oxidation was performed at 90°C (with temperature optimised at 60 and 90°C) and oxygen pressure kept at 8.5 bar using a glass-lined Parr reactor (model 4563), in batch mode under agitation with stirrer speed kept constant at 1000 rpm, using 0.5 g of catalyst for 10 g of glycerol dissolved in 90 g of de-ionized water. To this solution, NaOH pellets were added such that the mole ratio of glycerol to the base was always 1:2. In-depth details are provided elsewhere [14].

2.4. Catalyst characterisation

2.4.1. High resolution transmission electron microscopy (HRTEM)

HRTEM analysis was conducted on a field emission microscope, the JEM2100F electron microscope from JEOL Ltd., fitted with energy-dispersive X-ray spectroscopy (EDX), wavelength dispersive spectroscopy (WDS) and electron beam backscattered diffraction (EBSD) operating on Oxford Instruments software. The instrument was operated at an accelerating electron beam of 200 kV and images captured in the bright field mode. The Nano-measurer 1.2 'Scion Imager' software was used for particle size analysis.

2.4.2. Temperature programmed desorption (TPD)

The acidity of the materials was qualitatively measured by temperature programmed desorption (TPD) of NH_3 on a micromeritics automated catalyst characterisation system: model AutoChem II 2920 chemisorption analyser. NH_3-TPD studies were performed by loading 0.25 g catalyst in a U-tube reactor and cleaning the sample in a gas stream of helium at 120°C for an hour at a ramping rate of 10°C min^{-1} to remove moisture and other adsorbed species. A mixture of 10% NH_3 balanced in He was flushed over the sample isothermally at 120°C. After adsorption was achieved at 120°C, the NH_3 desorption measurements ensued at 120°C using the thermo-conductivity detector (TCD) and data collected up to 500°C at a ramping rate of 15°C min^{-1}.

2.5. Models and computational methods

2.5.1. Kinetic parameter estimation

The regression analysis of the experimental data was performed by use of Easy Regression Analysis (ERA 3.0) software [15, 16]. The software uses the sum of the square of residual deviations as the objective function. All the kinetic parameters were estimated at a 95% confidence limit using a modified adaptive random search algorithm.

2.5.2. DFT methodology

Unless otherwise stated, all electronic energies of reactants, products and transition states were determined by density functional theory (DFT) using the $DMol^3$ code [17, 18] within the

BIOVIA Materials Studio 2016 environment using the generalised gradient approximation (GGA), with a double numerical basis set (DNP) and the Perdew-Becker-Ernzerhof (PBE) exchange-correlation functional. All electrons were included in the calculations with unrestricted spin-polarization. A fine integration grid was used together with a Fermi smearing of 0.005 Hartree (Ha). The energy convergence tolerance was set to 1.0×10^{-5} Ha; the maximum force was 0.002 Ha/Å and maximum displacement was set at 0.005 Å. The self-consistent field (SCF) density convergence was set 1.0e − 6.

The complete linear synchronous transit and quadratic synchronous transit (LST/QST) method [19] was used to locate the transition state structures according to the optimised structures of reactants and products. The nudged elastic band (NEB) method [20], as implemented in $DMol^3$, was used to confirm that the transition state structures lead to the expected reactant and product molecular structures. Frequency calculations were performed to confirm the nature of all stationary points as either minima or transition states (TSs).

When investigating a reaction that takes place on the surface of a heterogeneous catalyst by quantum chemical methods, one of the main difficulties is modelling an infinite catalyst system as highlighted by Handzlik and Ogonowski [21], and a viable approach with the capacity of solving this problem is by the application of cluster models. Song et al. [22] claim that, ideally, cluster models are appropriate if a suitable boundary condition is obtained such that charges are in reasonable distribution on the surface. In this study, we chose a Al_3MoO_7H cluster to represent an active site of Mo on $MoO_3/\gamma\text{-}Al_2O_3$. Assuming that the oxidation states of the elements in the model are distributed as follows: $Al = 3^+$, $Mo = 4^+$, $H = 1^+$ and $O = 2^-$; then the overall charge of the cluster would be zero. From this point of view it can be assumed that when the substrate adsorbs on Al_3MoO_7H, it occupies the vacant sites on Mo to form a cluster model in which the Mo is in an approximate octahedral environment with a 6^+ oxidation state. This adsorption mode is consistent with the work of Kong et al. [23] who indicated how LAC could be formed from adsorbed intermediates such as pyruvaldehyde (PA).

3. Results

3.1. Catalyst characterisation

The sample of the NH_3-TPD profiles is displayed in **Figure 1** shows that the $Au\text{-}MoO_3/\gamma\text{-}Al_2O_3$ catalyst was more acidic than the original $\gamma\text{-}Al_2O_3$ support. All the materials displayed noticeable Lewis acidity (that is, electron accepting sites as opposed to the Brønsted acidity, which are regarded as proton donating sites) since the ammonia desorption occurred at the lower temperatures (below 300°C). The $\gamma\text{-}Al_2O_3$ support indicated Lewis acidity having different strengths with the weaker sites desorbing NH_3 at about 180°C while the stronger sites desorbed NH_3 at about 300°C.

On the addition of gold, the $Au/\gamma\text{-}Al_2O_3$ catalyst showed a shift in the two peaks to the higher temperatures, that is, 220 and 450°C, respectively. However, the addition of MoO_3 on $\gamma\text{-}Al_2O_3$ support exhibited only the (weaker) Lewis acid sites that desorbed NH_3 at the lower temperatures, peaking at 220°C. This finding is in agreement with a number of researchers who have

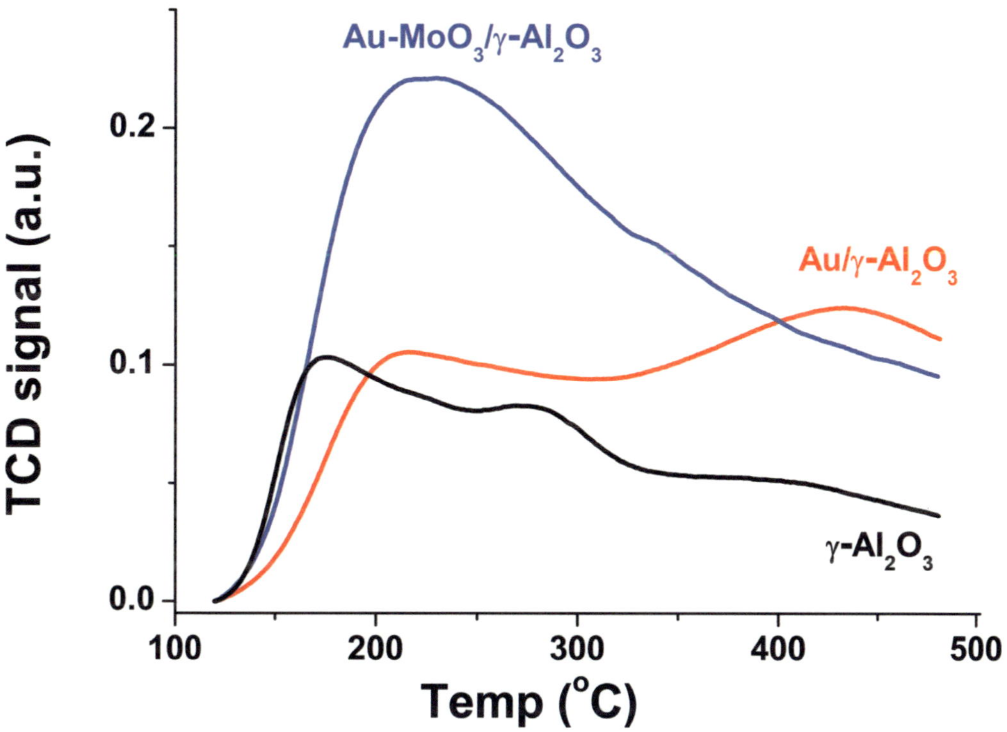

Figure 1. Qualitative analysis of the catalyst's acidity by the NH_3-TPD method.

recorded the absence of Brønsted acid sites in similar alumina systems, results which were further confirmed by their infrared studies, for example, by Lianecki et al. [24] and by XRD studies by Heracleous et al. [25]. It has been proposed by Gong et al. [26] that as long as the loading of MoO_3 on γ-Al_2O_3 is less than 16% (w/w), the nature of surface acid-base sites that exist on the surface would be predominantly of the Lewis type.

3.2. Reaction control: kinetic studies

3.2.1. *The effect of base as a reaction initiator*

Catalyst testing commenced with investigating the effect of NaOH as a reaction initiator on the kinetics of glycerol oxidation using Mintek's 0.9-wt% Au/γ-Al_2O_3 (AUROlite^{TM}) commercial catalyst. **Figure 2** displays a plot depicting glycerol conversion as a function of time indicating that higher base concentrations led to greater glycerol conversions thereby shortening reaction time. It has indeed been previously reported that the base acts an initiator for this reaction [5].

3.2.2. *Mass transfer limitations*

In order to optimise the reaction conditions for mass transfer, a number of influencing parameters, e.g. stirring speed, oxygen partial pressure, amount of catalyst and the initial

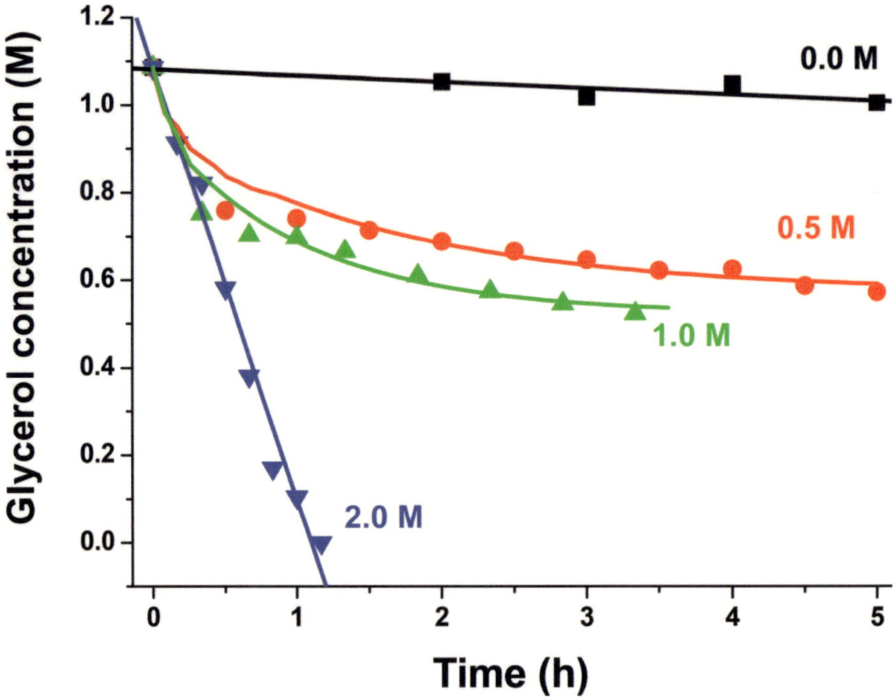

Figure 2. The effect of base concentration as a reaction initiator on glycerol oxidation using 0.5 g of 0.9-wt% Au/γ-Al_2O_3 catalyst at 90°C and O_2 pressure of 8.5 bar.

educt concentrations, were varied to arrive at the kinetic regime. The commercial 0.9-wt% Au/γ-Al_2O_3 (Aurolite^{TM}) catalyst was used to establish the mass transfer regime. The stirring rate was fixed at 1000 rpm and the amount of catalyst varied as shown in **Figure 3**. In theory, if the rate doubles with the doubling of the weight of the catalyst, then the reaction is controlled by kinetics; if this is not the case, then the reaction is controlled by mass transfer. The curve in **Figure 3** indicates that, under the defined experimental conditions, the ideal amount of catalyst necessary to achieve kinetic control was between 0.5 and 1.2 g. This 'reaction-limited' situation was ideal for the determination of intrinsic reaction kinetic parameters.

3.2.3. The effect of temperature on activation energy

In this work, the activation energy for the oxidation of glycerol over Au/γ-Al_2O_3 catalyst was experimentally determined. For a zero-order reaction, it can be shown that fractional conversion is linearly dependent on time and temperature through the relationships shown in Eq. (1) and Eq. (2):

$$X_A = \left(\frac{k}{C_{Ao}}\right) \cdot t \tag{1}$$

where X_A is the fractional conversion of reactant A,

http://dx.doi.org/10.5772/intechopen.70485

C_{A0} is the initial concentration of A,

k is the rate constant with units as mol L^{-1} min^{-1} and

t is time in minutes,

and that, k is dependent on temperature,

$$\mathbf{k} = \mathbf{A}.\mathbf{e}^{-\left(\frac{E_A}{RT}\right)} \tag{2}$$

where k is the rate constant with units as mol L^{-1} h^{-1}

A is the pre-exponential factor, specific to this reaction

E_A is the activation energy of the reaction,

R is the gas constant 8.314 J mol^{-1} K^{-1} and

T is the temperature in Kelvin.

Figure 4 shows the plot of glycerol conversion as a function of time for experiments that were carried out at 60 and 90°C. Usually, k is regarded as the rate coefficient of the overall reaction, which is some measure of catalyst activity, but in essence, as shown by Eqs. (1) and (2), k is temperature-dependent, as well as concentration-dependent, usually dependent on the initial concentration of the reactant.

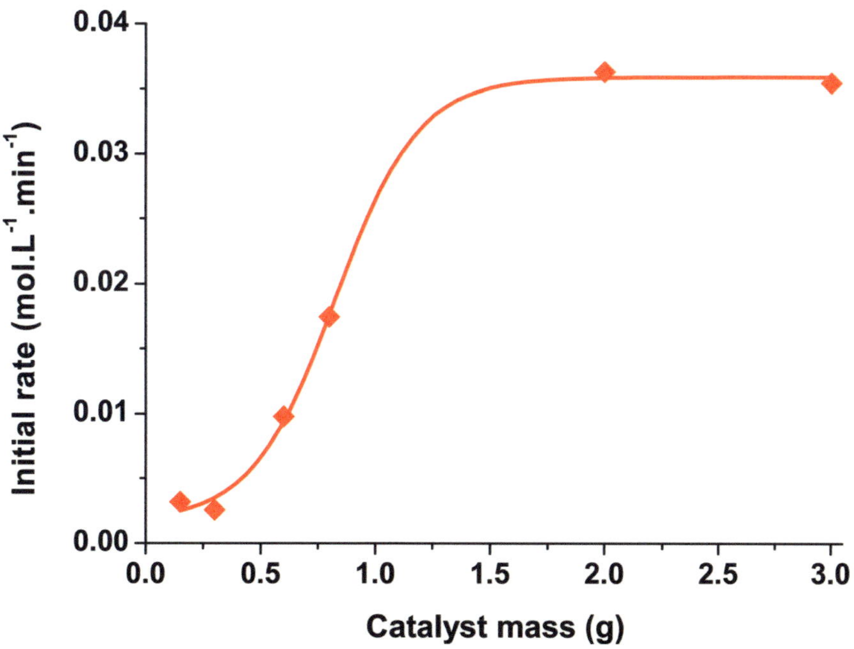

Figure 3. Initial rate as a function of catalyst mass for glycerol oxidation over 0.9-wt% Au/γ-Al_2O_3 (106–150 mμ): 1000 rpm, NaOH/glycerol = 2:1, at 60°C, under 8.5-bar O_2 pressure.

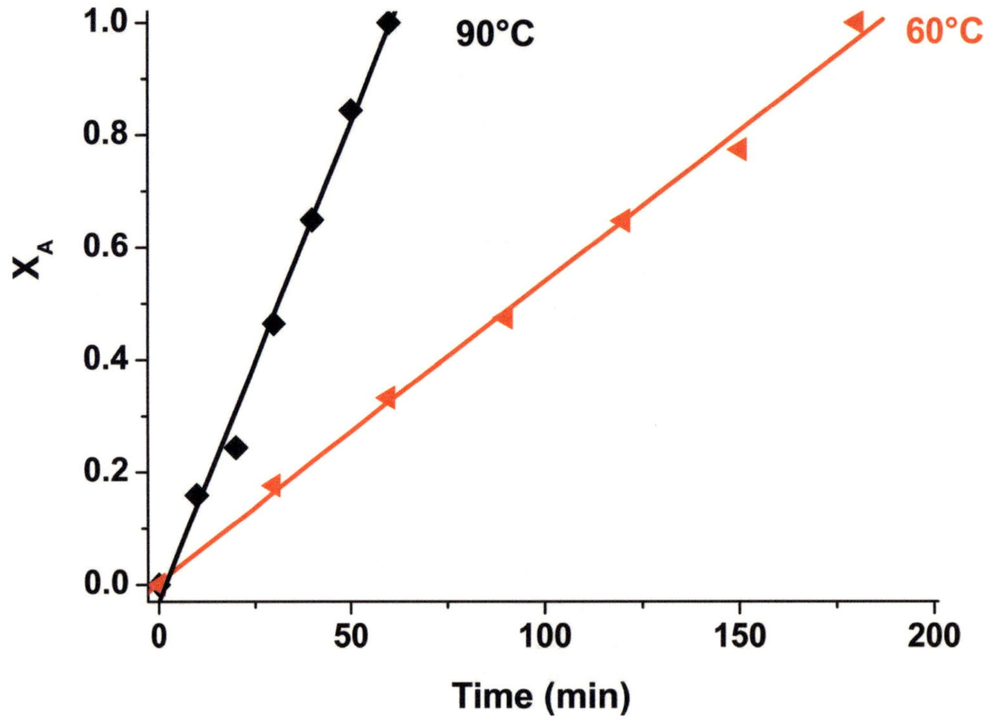

Figure 4. Conversion of glycerol over time by 0.9-wt% Au/γ-Al_2O_3 (106–150 mμ) catalyst; 1000 rpm, NaOH/glycerol = 2:1, at 60 and 90°C, under an O_2 pressure of 8.5 bar.

The rate constants, *k*, were estimated by non-linear regression of the model shown in Eq. (2) against experimental data and the results summarised in **Table 1**, showing the apparent E_A determined from the estimated rate constants using a two-point Arrhenius equation, thus:

$$E_A = \frac{T_1 \times T_2 \times R}{T_2 - T_2} ln\left(\frac{k_2}{k_1}\right) \tag{3}$$

By the use of a two-point Arrhenius equation with catalyst activity being measured at 60 and 90°C, the pre-exponential factor, *A*, was determined to be equal to 274 695 mol $dm^{-3}\,h^{-1}$ under the reaction conditions employed. In addition, the *k* values for the various catalysts were determined experimentally using Eq. (2). Then, since $k = 274695.e^{-E_A/RT}$, and in substituting the values for *A* and *k* in the Arrhenius equation, the E_A per catalyst were found to be in very close proximity, with the least active catalyst (Au/γ-Al_2O_3) having the highest E_A = 37.4 kJ mol^{-1}, followed by the Au-MoO_3/γ-Al_2O_3 with E_A = 35.4 kJ mol^{-1}. The rate constants (*k*) of the catalysts for overall glycerol conversion, calculated from experimental results were observed to be 2.22 and 1.14 mol $dm^{-3}\,h^{-1}$ for the Au-MoO_3/γ-Al_2O_3 and Au/γ-Al_2O_3, respectively, per gram of catalyst. Alternatively, when normalised to the amount of Au in the catalyst, the rate constants were found to be 642 mol $dm^{-3}\,h^{-1}\,g_{Au}^{-1}$ for Au-MoO_3/γ-Al_2O_3 catalyst and 252 mol $dm^{-3}\,h^{-1}\,g_{Au}^{-1}$ for Au/γ-Al_2O_3 catalyst.

Model[#]	Parameter	Temperature					
$X_A = \left(\frac{k}{C_{A0}}\right) \times t$	k		60°C			90°C	
		Value	Confidence limits		Value	Confidence limits	
		0.0059	0.0058	0.0061	0.018	0.017	0.019
$E_A = \frac{T_1 \times T_2 \times R}{T_2 - T_2} ln\left(\frac{k_2}{k_1}\right)$	E_A	Value			Confidence limits		
		37.4		36.1		38.1	

Parameter values estimated at 95% confidence level; k = mol L^{-1} min^{-1} and E_A = kJ mol^{-1}.
[#]R = 8.314 J K^{-1} mol^{-1}.

Table 1. Regression analysis of conversion vs. time data for the 0.9-wt% Au/γ-Al_2O_3 catalyst at different temperatures.

The experimentally obtained E_A in this work, as shown in **Table 1**, is in good agreement with the 38 kJ mol^{-1} value reported by Wörz et al. [27] for glycerol oxidation, although their work was based on a Pt-Bi/C catalyst. Demirel et al. [28] also reported activation energies of between 40 and 50 kJ mol^{-1} for the Au/C catalysed glycerol oxidation, which is a result not far from our own. Finally, Chornaja et al. [10] also reported an apparent E_A of about 39 ± 3 kJ mol^{-1} for glycerol oxidation over Pd/Al_2O_3, which is also consistent with our findings.

As a control, both the bulk γ-Al_2O_3 and MoO_3/γ-Al_2O_3 supports did not show any activity towards glycerol oxidation at 90°C.

3.2.4. The effect of Au nanoparticle size

The effect of Au particle size on the kinetics of glycerol conversion was also investigated. **Figure 5** shows TEM images of three Au/γ-Al_2O_3 catalysts with different Au particle sizes. When screened for glycerol oxidation under the same reaction conditions, the catalysts surprisingly revealed that the kinetics of this reaction strongly depend on the metal particle size, as shown in **Figure 6**. Big Au nanoparticles, *ca* 20 nm, exhibit first-order kinetics while the data for smaller Au nanoparticles, *ca* 4 nm, does not fit this model, instead showing zero-order kinetics as it has already been discussed. More details of this phenomenon have already been reported elsewhere [14].

3.3. Global kinetic model prediction

So far we have only concentrated on the depletion kinetics of the reactant. In this section, we explore the global kinetics of the reaction, i.e. we set-up a kinetic model that takes into account the full mass balance of the reaction. At full substrate penetration and surface coverage, conversion is not limited by mass transfer; it is a fixed quantity set by the zero-order kinetics. Accordingly, mass transfer terms have not been explicitly expressed in the kinetic model used in this study since kinetic data was collected under the kinetic control regime. Therefore, **Figure 7** presents the reaction network on which the kinetic modelling was based. The concentrations of all the compounds in the figure were taken into account in the calculations. Since the partial pressure (hence concentration) of O_2 was maintained high in excess, its surface coverage

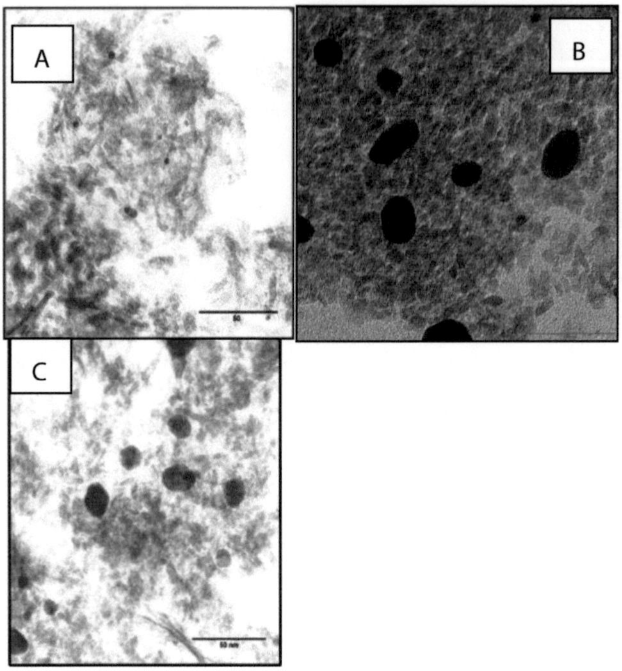

Figure 5. TEM images of $Au/\gamma\text{-}Al_2O_3$ catalysts prepared by various reducing agents: (A) reduced with 5% H_2 (~4 nm); (B) reduced with THPC (~17 nm); and (C) reduced with PVA–citrate (~21 nm). Catalyst preparation details were outlined elsewhere [14].

was assumed constant and the surface reactions were modelled as pseudo-monomolecular in the tested model.

The complete mass balance, based on **Figure 7**, is represented by Eqs. (4)–(8). The model is pseudo-zero-order overall and contains five parameters in total.

$$\frac{dC_1}{dt} = -(r1 + r3) = -k_1 \tag{4}$$

$$\frac{dC_2}{dt} = r1 - r2 = k_2 - k_3 \tag{5}$$

$$\frac{dC_3}{dt} = r2 = k_3 \tag{6}$$

$$\frac{dC_4}{dt} = r3 - r4 = k_4 - k_5 \tag{7}$$

$$\frac{dC_5}{dt} = r4 = k_5 \tag{8}$$

http://dx.doi.org/10.5772/intechopen.70485

Each rate constant, k_i, is defined as:

$$k_i = A_i e^{\frac{-E_{A_i}}{RT}} \tag{9}$$

where A_i is the frequency factor and E_{Ai} is the activation energy.

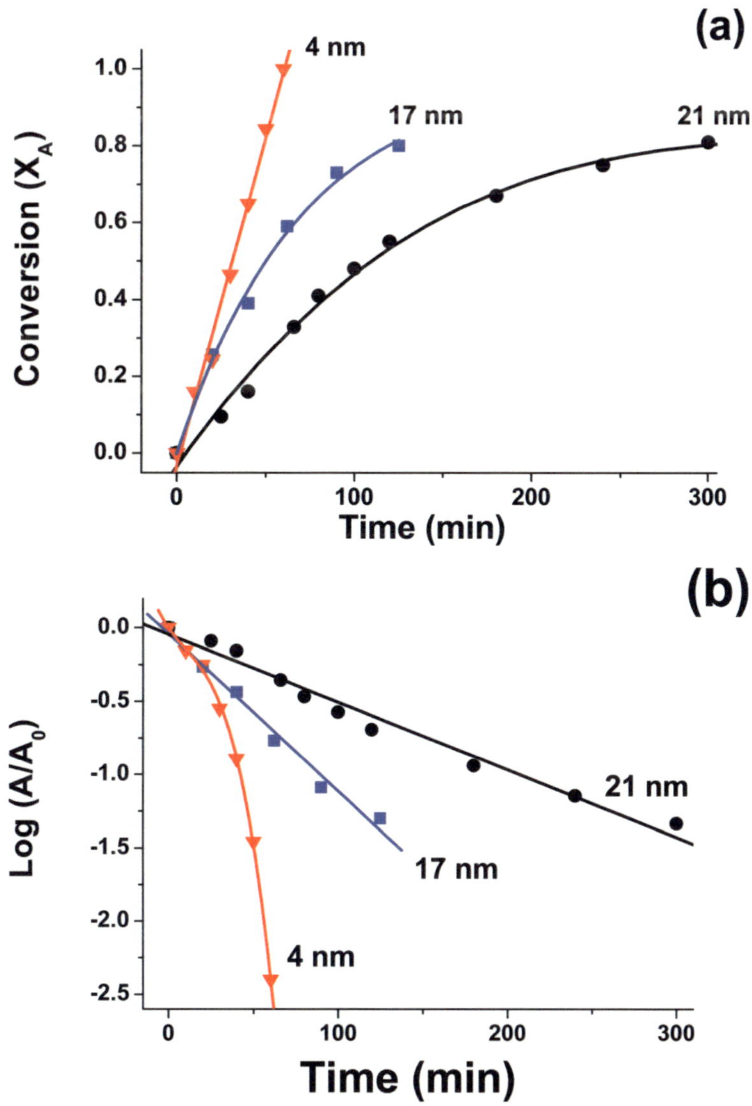

Figure 6. Glycerol consumption plots over Au/γ-Al_2O_3 catalysts prepared by various reducing agents: (a) Small gold nanoparticles (ca 4 nm) show zero-order kinetic behaviour by linearly fitting conversion as a function of reaction time; (b) big gold nanoparticles (ca 17 and 21 nm) show first-order kinetics by linearly fitting the log of concentration of glycerol as a function of reaction time. The catalysts were reduced with 5% H_2 (~4 nm), THPC (~17 nm) and PVA–citrate (~21 nm). Catalyst preparation details were outlined elsewhere [14].

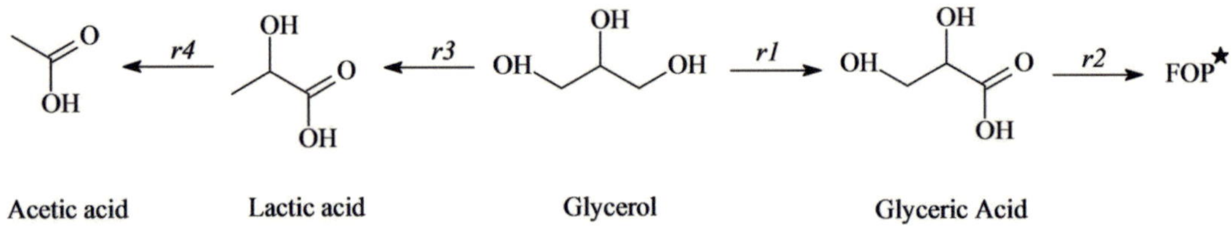

Figure 7. Reaction network used for the kinetic modelling of the glycerol oxidation. FOP* = further oxidation products (tartronic, oxalic, glycolic and formic acids).

The concentrations C_i were allocated as follows: C_1 = glycerol; C_2 = glyceric acid (GA); C_3 = FOP* (further oxidation products—FOP—were lumped together as tartronic acid, oxalic acid, glycolic acid and formic acid); C_4 = LAC and C_5 = acetic acid. The reason FOP were lumped together is that the study was only concerned with the apparent competition between the rate of formation of LAC and that of GA from glycerol and the possible effect of Lewis acidity on the rate of LAC formation. Generally, there was a good fit between the fitted pseudo-zero-order model and experiment for both tested catalysts. The results of the regression analyses of the model against experimental data showed a good fit as visually seen in **Figure 8**; the rate constants are summarised in **Table 2** for both Au/γ-Al_2O_3 and Au-MoO_3/γ-Al_2O_3, respectively. The estimated kinetic parameters were statistically significant.

From **Table 2**, one of the most intriguing results was the 'jump' in the rate of formation of LAC (k_4) over the Au-MoO_3/γ-Al_2O_3 catalyst relative to Au/γ-Al_2O_3. Indeed Eq. (10), which compares the rate of formation of LAC over the two catalysts, paints a clearer picture. It is conceivable that the extra Lewis acidity provided by Mo played a role in the promoted

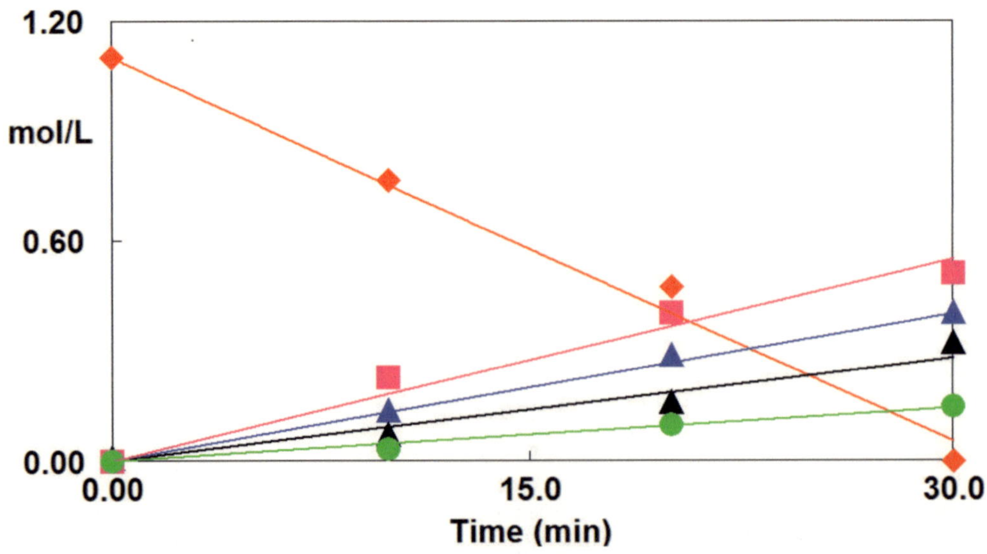

Figure 8. Comparison of computed and experimental data for glycerol oxidation assuming zero-order kinetics over Au-MoO_3/γ-Al_2O_3 (top) and Au/γ-Al_2O_3 (bottom); Glycerol (♦), glyceric acid (▲), FOP* (▲), LAC (■), acetic acid (●).

Parameter*	Catalyst					
	$Au/\gamma\text{-}Al_2O_3$			$Au/MoO_3\text{-}\gamma\text{-}Al_2O_3$		
	Value	Confidence limits		Value	Confidence limits	
k_1	0.019	0.018	0.02	0.035	0.032	0.038
k_2	0.021	0.019	0.023	0.023	0.018	0.028
k_3	0.008	0.007	0.01	0.009	0.006	0.013
k_4	0.004	0.002	0.006	0.023	0.019	0.028
k_5	0.001	0.000	0.003	0.005	0.002	0.008

*Parameter values were estimated at 95% confidence level; k = mol L^{-1} min^{-1}.

Table 2. Regression analysis of global kinetic model assuming pseudo-zero-order kinetics.

formation of LAC, as evidenced in Eq. (10) by the ratio of the rate constants of formation of LAC over the two catalysts.

$$\frac{k_{Au/MoO_3-Al_2O_3}}{k_{Au/Al_2O_3}} \cong 6 \tag{10}$$

As far as we are aware, no formation of LAC from Au-based catalysts (our work included) has been reported at 60°C except when Pd-based catalysts were used, which also gave very low turnover frequency (TOF) values (~60 h^{-1}) [10]. The TOF is the best measure of comparing catalytic performance as argued by Boudart [29]. TOF is regarded as the number of times that the overall catalytic reaction takes place per catalytic site per unit time for a fixed set of reaction conditions. Boudart has asserted that even though it may not be rigorous, a TOF measure leads to values that can be reproduced from one laboratory to another, besides catalysts of different characteristics being likened. We have therefore simplified TOF to mean the number of glycerol moles converted per the total number of moles of Au used per unit time of reaction, in hours, assuming that all the gold present in the catalyst was active. We have also assumed that the reaction occurred at constant temperature (90°C), O_2 pressure (8.5 bar), concentration (1.1 M of glycerol at t_0), and the ratio of reactants, while the extent of reaction was taken after half an hour. Thus, TOF was calculated by the equation:

$$\text{TOF} = \left[\frac{(\%\,\text{conversion}).(\text{initial moles (glycerol)})}{100}\right]\cdot\left[\frac{1}{\text{moles of Au}}\right]\cdot\left[\frac{1}{\text{time (h)}}\right] \tag{11}$$

For example, the TOF of the Aurolite™ commercial catalyst (0.9-wt% $Au/\gamma\text{-}Al_2O_3$) after half an hour was calculated to be 4.971 h^{-1}, compared to that of $Au\text{-}MoO_3/\gamma\text{-}Al_2O_3$ catalyst (*ca* 12.800 h^{-1}). This result shows that not only does the increased acidity due the presence of Mo enhance the selectivity to LAC, but it also drives the forward conversion of glycerol. This latter point could be related to gold-support interaction effects.

3.4. Proposed role of Mo^{n+} in lactic acid formation

It is generally accepted that LAC is formed from glycerol (GLY) via the intermediates as depicted in **Figure 9** [5, 30–32]. Under the reaction conditions employed in this study, dihydroxyacetone (DHA) was formed at 60°C and could be isolated; however, at 90°C, LAC was formed instead.

It is plausible that at the latter temperature, DHA still forms but reacts further with LAC, thus assuming the role of an (unstable) intermediate. In the presence of base, DHA can form from glycerol via the isomerisation of GLA [31]. Assary et al. [33] have reported that a Lewis acid-base pair catalyses this isomerisation reaction and it is conceivable that the enhanced formation of LAC over Au-MoO_3/γ-Al_2O_3 could have been a direct consequence of the increased Lewis acid-base pair sites on the catalyst. In this section, we apply transition state theory principles [34–36] via DFT simulations to extract thermodynamic and kinetic parameters for this isomerisation reaction. Experimentally, Rasrendra et al. [37] suggested that, under catalytic conditions, GLA can be successfully isomerised to LAC. The rest of this chapter discusses the potential role played by Mo in the formation of LAC over supported bi-functional Au catalysts at low temperatures, assuming the reaction occurs according to Eq. (12).

$$GLY \xrightarrow{Au} GLA \xrightarrow{Mo-OH} LAC \tag{12}$$

DFT calculations have been done only for the second part of the reaction, since the calculations for GLY to GLA over Au have been discussed in detail elsewhere [31, 38]. The mechanism shown in **Figure 10** proposes that a crucial part of the isomerisation of GLA to LAC could be an adsorbed pyruvaldehyde (PA) molecule on the molybdenum Lewis acid site forming a five membered ring C–C–O–Mo–O. The adsorbed PA is then hydrated by a surface hydroxyl, as advanced by Kong et al. [23]. The intermediate product then undergoes a hydride shift rearrangement. Finally, a proton is then transferred from a water molecule to

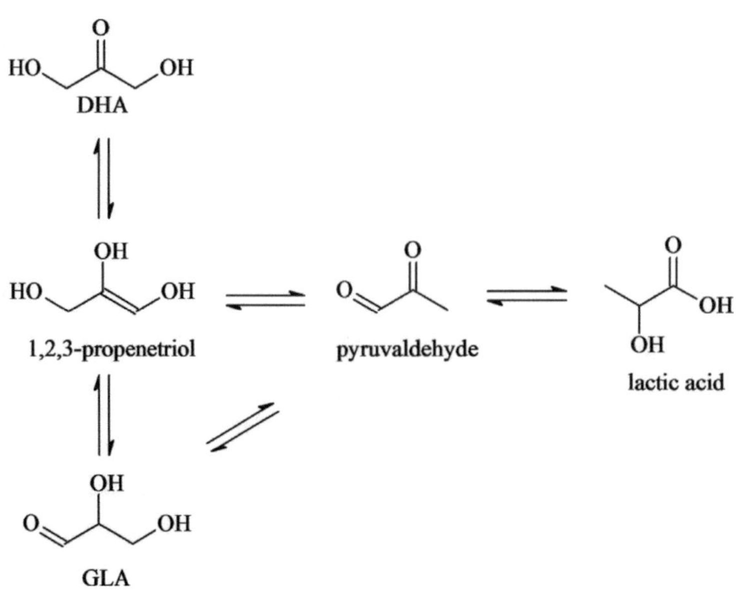

Figure 9. Proposed intermediates in the conversion of glycerol to LAC.

http://dx.doi.org/10.5772/intechopen.70485

complete the isomerisation and form a LAC molecule. Similar mechanisms have been proposed by numerous authors [5, 30, 33, 39–40], although none of these studies have modelled the mechanism at the DFT level of theory. The DFT predicted kinetic and thermodynamic parameters are summarised in **Table 3**. The reaction kinetics tool within BIOVIA Materials

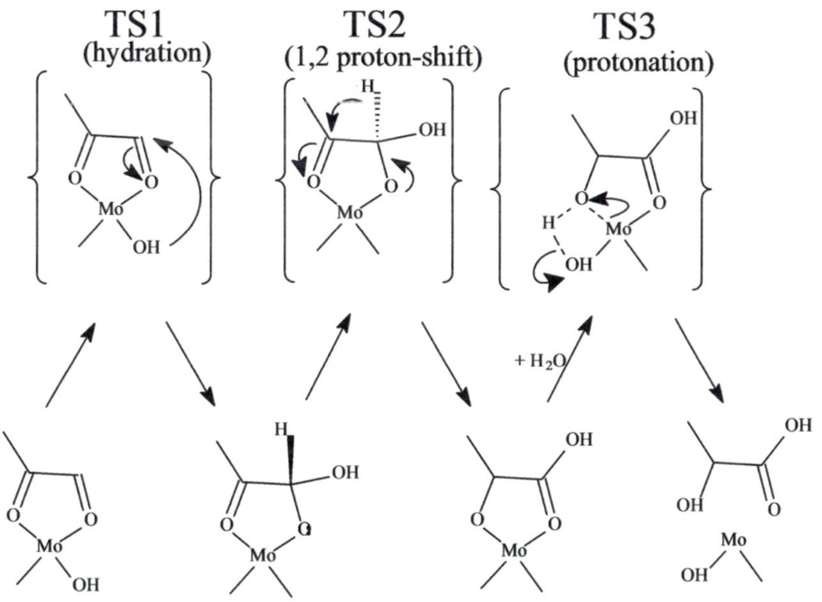

Figure 10. Possible Lewis acid-base pair dual site involved in the isomerisation of PA to LAC.

Transition state	Parameters (kJ mol^{-1})			
	Forward		Reverse	
	E_a	ΔH	E_a	ΔH
TS1 (hydration)	59.0	57.7	2.7	−57.7
TS2 (1,2 proton shift)	74.7	−57.0	131.7	57.0
TS3 (protonation)	229.7	−25.5	257.8	25.5

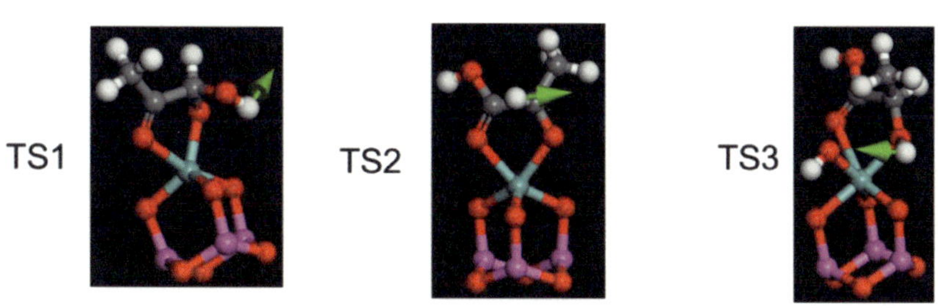

Table 3. DFT predicted kinetic and thermodynamic parameters for the Mo–OH catalysed isomerisation of PA to LAC. The direction of normal mode of vibration of the transition state is shown by the pointed arrow.

Studio 2016 was used to automate the calculations of all the parameters. Tunnelling corrections were included in all calculations. The protonation step is thermodynamically downhill, but has the highest activation barrier (229 kJ mol^{-1}) for the forward reaction as shown in the mentioned table. The magnitude of this barrier is in the order of the bond-dissociation energy of HO–H bond of a water molecule (H_2O) which requires about 268 kJ mol^{-1} at 298 K. Assuming that the reaction is controlled by kinetics, this step is the rate-limiting step. The high energy barrier for the rate-limiting step probably explains why LAC forms at 90°C under the reaction conditions employed, but none was observed at 60°C. A higher temperature is needed to overcome the barrier ($-E_a/R$) in order to get appreciable rates of the formation of the final product.

4. Conclusions

Both Au/γ-Al_2O_3 and Au-MoO_3/γ-Al_2O_3 showed zero-order kinetics under kinetic controlled glycerol oxidation. The apparent E_a of glycerol oxidation under these conditions was experimentally determined to be approximately 36 kJ mol^{-1}. Under the same reaction conditions, the presence of MoO_3 increased the formation of LAC sixfold over Au-MoO_3/γ-Al_2O_3 relative to Au/γ-Al_2O_3. A proposed DFT model suggested that protonation by an adsorbed water molecule might be the rate-limiting step in the isomerisation of PA to LAC catalysed by a Mo–OH Lewis acid-base pair.

Acknowledgements

The authors would like to thank both Mintek and Anglo-gold Ashanti through Project AuTek for granting the permission to publish this work and for financial support. In addition, we express gratitude to the South African Centre for High Performance Computing (CHPC) for their support in availing to us the infrastructure used in DFT modelling.

Author details

Thabang A. Ntho*, Pumeza Gqogqa and James L. Aluha

*Address all correspondence to: thabang.ntho@mintek.co.za

Mintek, Advanced Materials Division, Johannesburg, RSA

References

[1] Wee Y-J, Kim J-N, Ryu H-W. Biotechnological production of lactic acid and its recent applications. Food Technology Biotechnology. 2006;**44**(2):163-172

[2] Haruta M. Catalysis of gold nanoparticles deposited on metal oxides. CATTECH. 2002;**6**(3):102-115

[3] Ketchie WC, Fang Y, Wong MS, Murayama M, Davies RJ. Influence of gold particle size on the aqueous phase oxidation of carbon monoxide and glycerol. Journal of Catalysis. 2007;**250**:94-101

[4] Ketchie WC, Murayama M, Davies RJ. Selective oxidation of glycerol over carbon supported AuPd Catalysts. Journal of Catalysis. 2007;**250**:264-273

[5] Ketchie WC, Murayama M, Davies RJ. Promotional effect of hydroxyl on the aqueous phase oxidation of carbon monoxide and glycerol over supported Au catalysts. Topics in Catalysis. 2007;**44**(1–2):307-317

[6] Wang D, Villa A, Su D, Prati L, Schlogl R. Carbon supported gold nanocatalysts: Shape effect in the selective glycerol oxidation. ChemCatChem. 2017;**5**(9):2717-2723

[7] Villa A, Wang D, Su DS, Prati L. Gold sols as catalysts for glycerol oxidation: The role of stabilizer. ChemCatChem Catalysis. 2009;**1**:510-514

[8] Demirel S, Lucas M, Lehnert K, Claus P. Use of renewables for the production of chemicals: Glycerol oxidation over carbon supported gold catalysts. Applied Catalysis B: Environmental. 2007;**70**(1-4):637-643

[9] Royker M, Case J, van Steen E. Platinum promotion of Au/Al_2O_3 catalysts for glycerol oxidation: Activity, selectivity, and deactivation. The Journal of Southern African Institute of Mining and Metallurgy. 2012;**7A**:577-581

[10] Chornaja S, Dubencov K, Kampars V, Stepanova O, Zhizhkun S, Serga V, et al. Oxidation of glycerol with oxygen in alkaline aqueous solutions in the presence of supported palladium catalysts prepared by extractive pyrolytic method. Reaction Kinetic Mechanism Catalysis. 2013;**108**:341-357

[11] Carretin S, McMorn P, Johnston P, Griffin K, Hutchings GJ. Oxidation of glycerol using supported Pt, Pd and Au catalysts. Physical Chemistry Chemical Physics. 2003;**5**:1329-1336

[12] Stobbe-Kreemers AW, Vanleerdam GCV, Jacobs JP, Brogersma HH, Scholten JJF. Characterization of γ-alumina-supported vanadium oxide monolayers. Journal of Catalysis. 1995;**152**:130-136

[13] Raubenheimer HG, Cronje S. Progress in Chemistry, Biochemistry and Biotechnology. Chichester: John Wiley & Sons; 1999. 573-576

[14] Ntho T, Aluha J, Gqogqa P, Raphulu M, Pattrick G. $Au/\gamma\text{-}Al_2O_3$ catalysts for glycerol oxidation: The effect of support acidity and gold particle size. Reaction Kinetic Mechanism and Catalysis. 2013;**109**(1):133-148

[15] Belohlav Z, Zamostny P, Kluson P, Volf J. Application of random-search algorithm for regression analysis of catalytic hydrogenations. Canadian Journal of Chemical Engineering. 1997;**75**:735-742

[16] Zamostny P, Belohlav Z. A software for regression analysis. Computers and Chemistry. 1999;**23**:479-485

[17] Delley B. An all-electron numerical method for solving the local density functional for polyatomic molecules. The Journal of Chemical Physics. 1990;**92**:508-517

[18] Delley B. From molecules to solids with the DMol3 approach. Journal of Chemical Physics. 2000;**113**:7756-7764

[19] Govind N, Petersen MM, Fitzgerald G, King-Smith D, Andzelm J. A generalized synchronous transit method for transition state location. Computational Materials Science. 2003;**28**(2):250-258

[20] Sheppard D, Terrell R, Henkelman G. Optimization methods for finding minimum energy paths. The Journal of Chemical Physics. 2008;**128**:134106-134116

[21] Handzlik J, Ogonowski J. Theoretical study on ethene metathesis proceeding on Mo^{VI} and Mo^{IV} methylidene centres of heterogeneous molybdena-alumina catalyst. Journal of Molecular Catalysis A: Chemical. 2001;**175**:215-225

[22] Song X, Liu G, Yu J, Rodrigues AE. Density functional theoretical study of water molecular adsorption on surface of MoO_3 with the cluster model. Journal of Molecular Structure (Theochem). 2004;**684**:81-85

[23] Kong L, Li G, Wang H, He W, Ling F. Hydrothermal catalytic conversion of biomass for lactic acid production. Journal of Chemical Technology and Biotechnology. 2008;**83**:383-388

[24] Lianecki M, Malecka-Grycz M, Domka F. Water-gas shift reaction over sulfided molybdenum catalysts I. Alumina, titania and zirconia-supported catalysts. Applied Catalysis A: General. 2000;**196**:293-303

[25] Heracleous E, Lemonidou AA, Lercher JA. Mechanistic features of the ethane oxidative dehydrogenation by in situ FTIR spectroscopy over a MoO_3/Al_2O_3 catalyst. Applied Catalysis A: General. 2004;**264**:73-80

[26] Gong J, Ma X, Wang S, Yang X, Wang G, Wen S. Effect of Mo content in $MoO_3/\text{-}\gamma\text{-}Al_2O_3$ on the catalytic activity for trans-esterification of dimethyl oxalate with phenol. Reaction Kinetics and Catalysis Letters. 2004;**83**(1):113-120

[27] Wörz N, Brandner A, Claus P. Platinum-bismuth catalyzed oxidation of glycerol: Kinetics and the origin of selective deactivation. Journal of Physical Chemistry C. 2010;**114**:1164-1172

[28] Demirel S, Lucas M, Warna J, Salmi T, Murzin D, Claus P. Reaction kinetics and modelling of the gold catalysed glycerol oxidation. Topics in Catalysis. 2007;**44**(1-2):299-305

[29] Boudart M. Turnover rates in heterogeneous catalysis. Chemical Reviews. 1995;**95**:661-666

[30] Shen Y, Zhang S, Li H, Ren Y, Liu H. Efficient synthesis of lactic acid by aerobic oxidation of glycerol on Au-Pt/TiO_2 catalysts. Chemistry – A European Journal. 2010;**16**:7368-7371

[31] Zope BN, Hibbits DD, Neurock M, Davis RJ. Reactivity of the gold/water interface during selective oxidation catalysis. Science. 2010;**330**:74-78

[32] Román-Leshkov Y, Davis ME. Activation of carbonyl containing molecules with solid Lewis acids in aqueous media. ACS Catalysis. 2011;**1**:1566-1580

[33] Assary RS, Curtiss LA. Theoretical study of 1, 2-hydride shift associated with the isomerisation of glyceraldehyde to dihydroxy acetone by Lewis acid active site models. Journal of Physical Chemistry A. 2011;**115**:8754-8760

[34] Evans MG, Polanyi M. Some applications of the transition state method to the calculation of reaction velocities, especially in solution. Transactions of the Faraday Society. 1935;**31**:875-894

[35] Eyring H. The activated complex in chemical reactions. Journal of Chemical Physics. 1935;**3**:107-116

[36] Geng Z, Zhang M, Yu Y. Theoretical investigation on pyrolysis mechanism of glycerol. Fuel. 2012;**93**:92-98

[37] Rasrendra CB, Marketihartha IGBN, Adisasmito S, Heeres HJ. Green chemicals from d-glucose: Systematic studies on catalytic effects of inorganic salts on the chemo-selectivity and yield in aqueous solutions. Topics in Catalysis. 2010;**53**(15-18):1241-1247

[38] Shang C, Liu Z-P. Origin and activity of gold nanoparticles as aerobic oxidation catalysts in aqueous solution. Journal of American Chemical Society. 2011;**133**:9938-9947

[39] West RM, Holm MS, Saravanamurugan S, Xiong J, Beversdorf Z, Christensen CH. Zeolite H-USY for the production of lactic acid and methyl lactate from C-3 sugars. Journal of Catalysis. 2010;**269**:122-130

[40] Rasrendra CB, Fachri BA, Marketihartha IGBN, Adisasmito S, Heers HJ. Catalyic conversion of dihydroxyacetone to lactic acid using metal salts in water. Journal of Sustainable Chemistry. 2011;**4**:768-777

Hydrothermal Precipitation of β-FeOOH Nanoparticles in Mixed Water/Alcohol Solvent

Mahabubur Chowdhury

Additional information is available at the end of the chapter

http://dx.doi.org/10.5772/intechopen.70503

Abstract

In this research, synthesis of β-FeOOH nanoparticles was carried out using different alcohol/water mixed solvents. Four different alcohols were mixed with water to form solution of different surface tension. A relationship between particle size and surface tension has been drawn from theoretical analysis. A linear relationship was shown to exist between surface tension and particle size under the reported conditions. Statistically designed experiments were conducted to evaluate the interactions of process parameters on the particle growth. A generic correlation has also been developed empirically to predict particle size.

Keywords: β-FeOOH, nanorod, crystal growth, surface tension, generic correlation

1. Introduction

Akaganeite, β-FeOOH, a type of iron oxy hydroxide has been studied intensively not only because of its great technological and scientific interest but also for many applications [1]. β-FeOOH is a promising electrode material that has potential in rechargeable batteries due to its hollandite-like crystal structure with tunnel-like cavities [2]. β-FeOOH has also found applications in hydroprocessing of coal, removal of arsenate/arsenite [3] and phosphate from water [4]. Due to these intriguing applications, numerous works have investigated the methods of akaganeite synthesis [2]. Surface tension of solvent plays an important role in the formation of 1D β-FeOOH nanorods. A low surface tension of solvent can promote the growth of β-FeOOH nanorods [5]. Addition of medium that can lower the surface tension of precursor solution can alter the thermodynamic properties of the reaction system and subsequently affect the nucleation kinetics, which would result in various morphology and particle sizes. The application of alcohol/water mixture during the synthesis of different metal oxide nanoparticles can be found in the literature. Submicrometer ZrO_2 nanomaterial was synthesised by Hu and Chen [6] using alcohol/water mixture. Fang and Chen [7] used a mixed solvent of water/n-propanol to synthesise titania powders. $ZrO_2(Y_2O_3)$ nanoparticles were synthesised by Li and Gao [8] using ethanol/water mixture. So far, the effect of mixed solvent properties on the growth of 1D β-FeOOH nanorods were not evaluated despite of its numerical industrial and scientific applications. In our previous work, we have demonstrated that alcohol played an important role in controlling nucleation rate and particle size of 1D β-FeOOH nanoparticles [9, 10]. Previously reported work in the literature considered the dielectric constant, ε, properties of the water/alcohol mixture to be of significance in controlling particle growth, discarding the effect of surface tension, γ. However, it is interesting to note that the Coulomb interaction is very weak in water and solvent with high ε [11] and Coulomb interaction is directly related to ε of the solvent. This simple phenomenon makes the application of the ε of solvent only to relate to particle growths as described in the literature susceptible.

The application of mixed solvent is a new approach in the synthesis and processing of materials [12]. However, the role of surface tension is neglected when relating particle growth with solvent properties. Tuning the surface tension of the water/alcohol can affect the colloidal interaction between solid particles. Most of the literature reported on the use of alcohol/water mixture on the precipitation of submicro meter-sized TiO_2, ZrO_2, SiO_2 and CeO_2 particles. Published literature provides evidence that the formation of nanoparticles strongly related to the solution pH, precursor salt concentration, time and temperature, etc. However, very few attempts have been made in the literature to describe the interaction between the process parameters from statistically designed experiments. Moreover, only one work was done to describe the effect of process variables on particle size [13]. However, the empirical correlation reported can only predict spherical shape particles and the experiments were not designed statistically to evaluate the actual relationship between process variables.

Development of mathematical relationship that combines fundamental properties and empirical quantities derived from statistical analysis can be very useful in understanding the actual effect of process interactions on the particle growth. In a previous work [9], mathematical relation between solvent surface tension and particle growth was reported. However, the effect of various process parameters on the particle growth was not accounted in that model. Hence, in this chapter, an evaluation of the fundamental relationship between solvent surface tension, precursor concentration, time and β-FeOOH nanorod growth is presented via statistically designed experiments.

According to classical electrostatic model, the chemical potential of two phases are equal in equilibrium and can be written as [11, 12]:

$$\mu^o{}_s + KT\ln C_S = \mu^o{}_L + KT\ln C_L \tag{1}$$

where μ^o is the standard chemical potential, C is the concentration of solute (S and L represent solid and liquid phases, respectively), T is the temperature expressed in Kelvin and K is Boltzmann constant. Eq. (2) also known as Coulombs interaction can be used to express the energy required to separate the charged ions from the solid state [11, 12]:

$$\Delta\mu^o = \frac{Z_+ Z_- e^2}{4\pi\varepsilon_o\varepsilon(r_+ + r_-)} \tag{2}$$

where ε_o represents the permittivity in vacuum and ε is dielectric constant of a given solution. r_+ and r_- present the radii of ions charged. Based on Eqs. (1) and (2), Chen and Chang [12] proposed the below equation:

$$\frac{2\gamma}{rkT\rho} = \ln C + \frac{Z_+ Z_- e^2}{4\pi\varepsilon_o\varepsilon KT(r_+ + r_-)} \tag{3}$$

where m is the weight of the solute molecule; γ is the interfacial energy in J/m^2, the term interfacial energy can also be written as nm^{-1}; r is the particle size of a spherical particle and C is the solute concentration. The author linearized Eq. (3) in terms of particle size r(Y) and dielectric constant ε(X). It was assumed that only dielectric constant is significant and other properties are insignificant. However, according to Israelachvili [11], Coulombic interaction is much weakened in water. So, it can be postulated that, for homologous mixture of water and alcohol, the Coulombic interaction will be very weak. Hence, it can also be hypothesised that the surface tension of solvent will directly affect the nucleation and growth of particles.

2. Experimental

2.1. Materials and method

Analytical-grade $FeCl_3{\cdot}6H_2O$ and NH_4OH (B & M Scientific, Cape Town, R.S.A) were used as they are without any further purification. Deionised water and alcohol (methanol, ethanol and propanol) were mixed together with different ratios to vary the surface tension of solvent. Ammonium hydroxide, NH_4OH, was added drop wise until the mixture pH reached a value of 10. A certain amount of $FeCl_3{\cdot}6H_2O$ was added to the solution to make up 0.05 M solution (unless stated otherwise), and the solution was stirred until the iron salt was dissolved. The pH of the final solution (prior to heating) was recorded. The pH was always kept constant at ~2 to isolate the effect of pH on the particle growth in each case. The solution was replaced in a teflon-lined pressure autoclave. The temperature was 100°C. A reaction time of 2 h was used to synthesise the particles.

3. Results and discussion

3.1. Correlation between particle size and alcohol surface tension

Three different alcohols (methanol, ethanol and propanol) were used to tune the surface tension of the solvent. If the hypothesis conceived earlier is true, then there will be a qualitative relationship between particle size, D_e, and surface tension of solvent, γ. The particles synthesised in this study were rod-shaped particles. The length and diameter of the particles were measured from TEM images. A minimum number of 250 particles was counted for

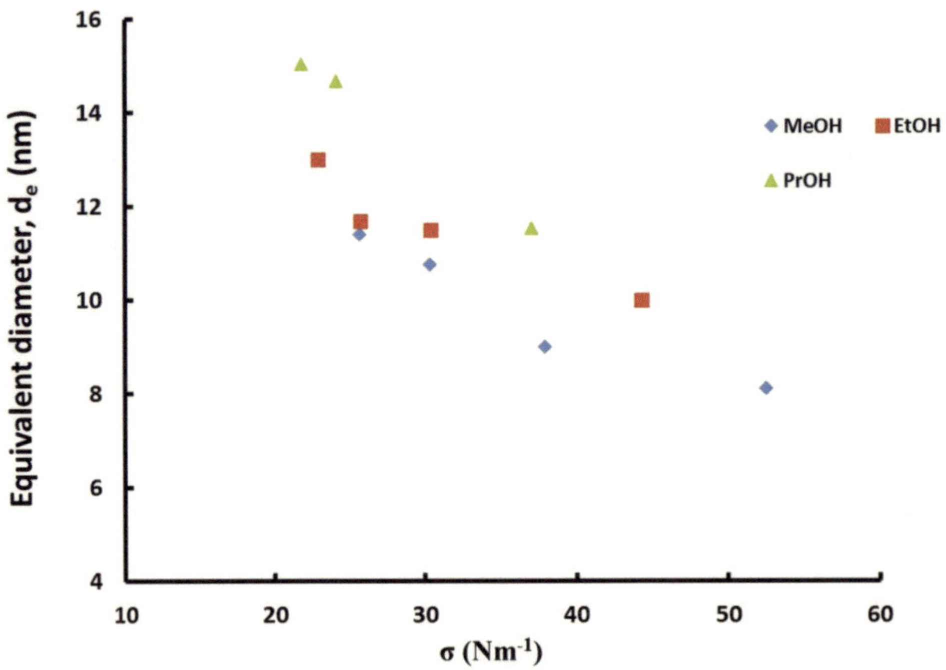

Figure 1. Correlation between particle size and solvent surface tension.

statistical purpose. The equivalent diameter of the rod-shaped particles was calculated using Huebscher formula to evaluate the role of surface tension on particle size [14]:

$$D_e = 1.30 \times \left[\frac{(a \times b)^{0.625}}{(a + b)^{0.25}}\right] \tag{4}$$

where a and b are the measured diameter and length of the particles in this study. It can be seen from **Figure 1** that there is a linear relationship that holds approximately independent of the nature of the alcohol used to synthesise the particles. This illustration proves the point that the surface tension of the mixed solvent can also be used to relate nucleation and growth.

3.2. A generic correlation to predict particle growth

Statistically designed experiments are effective optimising tools where the process is influenced by various external factors. In this chapter, a two level three factor factorial design was used to evaluate the effect of interaction of various parameters that have been manipulated during the synthesis of β-FeOOH particles. Previous work [9, 10] has shown that the particle phase is very sensitive to reaction temperatures and pH. Hence, reaction temperature and pH

Factors	Low level (−)	High level (+)
A: $FeCl_3$ concentration [M]	0.05	0.5
B: % solvent to water ratio	30	90
C: Time (h)	2	12

Table 1. Real amount of each factor used in the factorial trial experiments.

were omitted from the factorial trial for simplicity sake. Furthermore, butanol was used as solvent to synthesise β-FeOOH particles to validate the relations between particle growth and solvent surface tension. **Table 1** presents the real amount of each parameters that have been used at low and high levels, which is assigned by a positive (+) and negative (−) sign, respectively.

3.3. Evaluation of the effects of factors and the interaction between the factors on particle growth

Table 2 presents the obtained equivalent diameter at different experimental conditions according to the two level three factor factorial design. A Pareto chart is presented in **Figure 2** to assess the interaction between process parameters and obtained equivalent diameter of the particles. The factorial design can cover the main and interaction effects of the parameters within the whole range of selected parameters. Evaluation of the effect of principal factors revealed that these parameters have positive effects on the obtained equivalent diameters of the particles.

Based on the significance of effects from **Figure 2**, a generalised empirical correlation that takes solvent surface tension, precursor concentration and time into account is proposed:

Samples	$FeCl_3$ concentration [M]	% alcohol to water ratio used as solvent	Time (h)	D_e
1	0.5	90	2	17
2	0.05	30	12	17
3	0.5	30	12	90
4	0.05	30	2	16
5	0.5	90	12	43
6	0.05	90	12	16
7	0.05	90	2	15
8	0.50	90	2	7

Table 2. Results obtained for a two level factorial design.

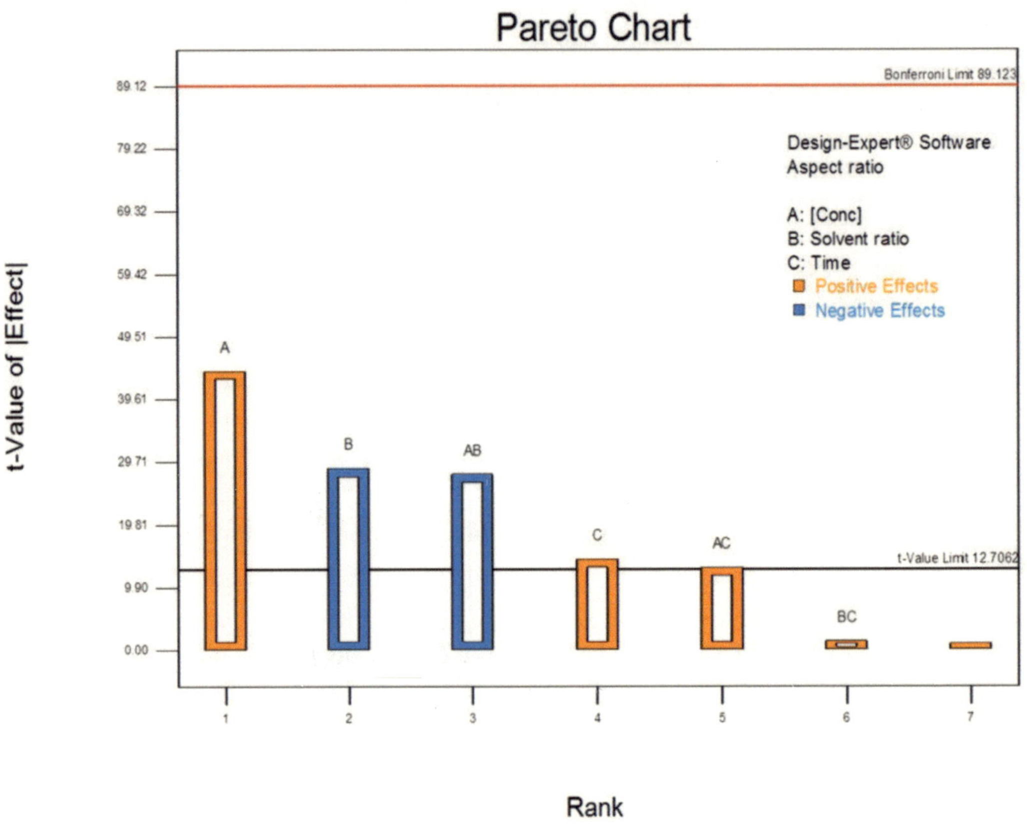

Figure 2. Estimated effects of factors on particle aspect ratio using factorial design.

$$Y = \beta_o\gamma + \sum_{i=A}\beta_i X_i + \sum_{i=A}\sum_{j=A\neq i}\beta_{ij}X_iX_j \tag{5}$$

where Y is the predicted response (D_e in this case), X_i is the un-coded or coded values of the factors ($FeCl_3$ concentration denoted by A, % alcohol to water ratio is denoted by B and time is denoted by C), β_o and β_i are constant and γ is the surface tension of alcohol used. The corresponding response model for the obtained aspect ratio of the particles, which are valid for un-coded factor, is:

$$\mathrm{Y} = 0.39\gamma + \beta i[A] + 0.047[B] - 0.38856[C] - 1.8[A^*B] + 5.05[A^*C] + 0.004[B^*C] \tag{6}$$

Linear regression can be used to obtain corresponding values of β_i. **Figure 3** presents typical comparison of experimental and predicted values using different β_i values. It can be seen from **Figure 3** that the comparison between experimental and predicted values agrees very well.

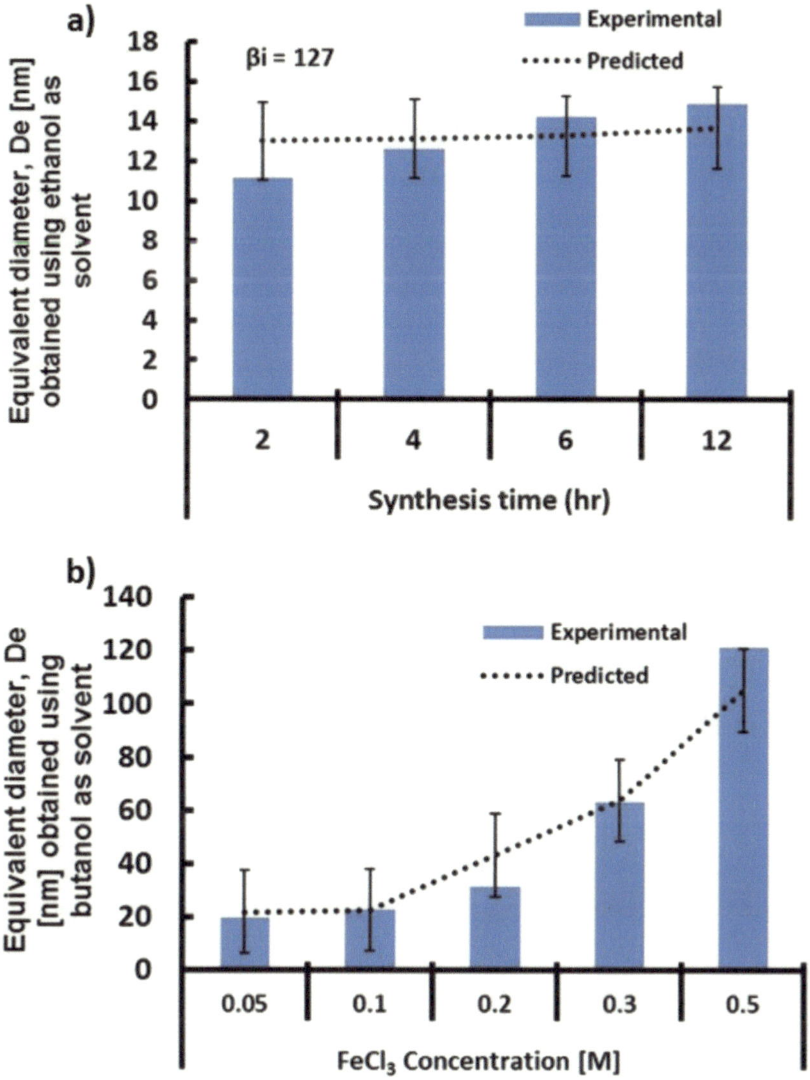

Figure 3. Typical comparison between experimental data and predicted value using different β_i constants for ethanol (a) and (b) butanol solvents.

4. Conclusion

Interaction between solvent surface tension and various process parameters on the synthesis of different β-FeOOH morphology has been evaluated for the first time assuming the presence of low Coulombic interaction. A linear relationship was found between particle size and surface tension of the solvent. Statistically designed experiments were performed to evaluate the interaction between particle growth and process parameters. A generic correlation has been developed to predict particle growth. The correlation can be extended further to predict particle size of other type of materials. The results obtained in this work clearly indicate that the application of only dielectric constant to relate particle nucleation and growth is not adequate. A combination of surface tension and dielectric constant together will be appropriate.

Acknowledgements

The author would like to thank National Research Foundation of South Africa (Grant no: 88220) for financial support.

Conflicts of interest

The author declares no conflicts of interest.

Author details

Mahabubur Chowdhury[1,2]*

*Address all correspondence to: chowdhurym@cput.ac.za

1 Flow Process and Rheology Centre, Cape Peninsula University of Technology, Cape Town, South Africa

2 Department of Chemical Engineering, Cape Peninsula University of Technology, Cape Town, South Africa

References

[1] Wei C, Nan Z. Effects of experimental conditions on one-dimensional single-crystal nanostructure of β-FeOOH. Materials Chemistry and Physics. 2011;**127**:220-226

[2] Chen M, Jiang J, Zhou X, Diao G. Preparation of akaganeite nanorods and their transformation to sphere shape hematite. Journal of Nanoscience and Nanotechnology. 2008;**8**: 3942-3948

[3] Chen L, Yang X, Chen J, Liu J, Wu H, Zhan H, Liang C, Wu M. Continuous shape- and spectroscopy-tuning of hematite nanocrystals. Inorganic Chemistry. 2010;**49**:8411-8420

[4] Kuang D, Xu A, Fang Y, Liu H, Frommen C, Fenske D. Surfactant-assisted growth of novel PbS dendritic nanostructures via facile hydrothermal process. Advanced Materials. 2003;**15**:1747-1750

[5] Shao H-F, Qian X-F, Yin J, Zhu Z-K. Controlled morphology synthesis of β-FeOOH and the phase transition to Fe_2O_3. Journal of Solid State Chemistry. 2005;**178**:3130-3136

[6] Hu Y, Chen K. Crystal splitting in the growth of β-FeO(OH). Journal of Crystal Growth. 2007;**308**:185-188

[7] Fang C-S, Chen Y-W. Preparation of titania particles by thermal hydrolysis of $TiCl_4$ in n-propanol solution. Materials Chemistry and Physics. 2003;**78**:739-745

[8] Li W, Gao L. Nano ZrO_2 (Y_2O_3) particles processing by heating of ethanol–aqueous salt solutions. Ceramics International. 2001;**27**:543-546

[9] Chowdhury M, Fester V, Kale G. Growth kinetics evaluation of hydrothermally synthesized β-FeOOH nanorods. Journal of Crystal Growth. 2014;**387**:57-65

[10] Chowdhury M, Fester V, Kale G, Cespedes O. Hydrothermal precipitation of β-FeOOH nanostructure(s) in mixed solvent: Study of their morphological and structural evolution. Journal of Nanoparticle Research. 2014;**16**:1-11

[11] Israelachvili JN. 3—Strong Intermolecular Forces: Covalent and Coulomb Interactions, Intermolecular and Surface Forces. 3rd ed. San Diego: Academic Press; 2011. pp. 53-70

[12] Chen H-I, Chang H-Y. Homogeneous precipitation of cerium dioxide nanoparticles in alcohol/water mixed solvents. Colloids and Surfaces A: Physicochemical and Engineering Aspects. 2004;**242**:61-69

[13] Chiu C-A, Hristovski KD, Dockery R, Doudrick K, Westerhoff P. Modeling temperature and reaction time impacts on hematite nanoparticle size during forced hydrolysis of ferric chloride. Chemical Engineering Journal. 2012;**210**:357-362

[14] Yuan Q, Aryanti N, Hou R, Williams RA. Performance of slotted pores in particle manufacture using rotating membrane emulsification. Particuology. 2009;**7**:114-120

Chapter 9

Adsorption, Kinetics and Photoactivity of ZnO-Supported Fly Ash-Sepiolite Ternary Catalyst

Ayşe Neren Ökte

Additional information is available at the end of the chapter

http://dx.doi.org/10.5772/intechopen.70504

Abstract

Nanocomposites have been attracting more attention in various fields. In this chapter, ZnO-supported fly ash-sepiolite (ZnO-FA-Sep) was prepared as a ternary composite for the evaluation of adsorption capacities and photocatalytic activities. Characterization studies supplied information about the surface morphology variation before and after ZnO loading within the FA-Sep environment. Strong dark adsorption capacities of the supported catalysts improved their photocatalytic performances, in terms of methyl orange (MO) decolorization and degradation processes. This study not only provided important inspirations for developing support materials but also opened new features to facilitate the photocatalyts' performances.

Keywords: supported catalyst, methyl orange, adsorption, kinetics, photocatalysis

1. Introduction

Adsorption has been used extensively in industrial processes for separation and purification. Recent research has been focused on the development of low cost adsorbents. Among many alternatives, fly ash cost-effectively improves the performance of products it is added to [1]. Although, significant quantities are being used in a range of applications, it mainly works in tandem with cement in the production of concrete products. However, there is still large amount of fly ash production. An efficient usage of fly ash creates positive environmental impacts. Fly ash use conserves natural resources and avoids landfill disposal of ash products. By making concrete more durable, life cycle costs of roads and structures are reduced. Furthermore, fly ash use partially displaces production of other concrete ingredients, resulting in significant energy savings and reductions in water consumption and greenhouse gas emissions.

Sepiolite is a natural hydrated magnesium silicate fibrous clay with a composition of $Si_{12}Mg_8O_{30}(OH)_4(OH_2)_4 \cdot 8H_2O$. It exhibits high surface area, porosity, good chemical stability and mechanical behavior [2, 3]. Its applications focused on the use as additives and adsorbents. The crystalline structure is composed of alternating silicate blocks and cavities. The presence of silanol groups (Si-OH) makes use of sepiolite as a support for metals and metal oxide nanoparticles [4].

ZnO is one of the most promising photocatalysts under the ultraviolet radiation to protect the environment by degradation of organic pollutants in water and air. In general, a photocatalytic reaction starts with the generation of electrons and holes by photoexcitation. Then, these charge carriers migrate to the surface of the photocatalyst and react with adsorbed electron acceptors and donors, respectively. Thus, an efficient photocatalyst requires a suitable band-gap, facile separation and transportation of electrons and holes for the feasibility of potential redox reactions. ZnO with its high photo sensitivity and large bandgap plays a significant role for reduction and oxidation processes. However, usage of ZnO nanoparticles in catalytic slurries alone creates fast charge carrier recombination and requires long-time centrifugation for the removal process. An effective way to overcome these difficulties is to immobilize the ZnO nanoparticles on the inner and outer surfaces of inorganic porous supports with the formation of nanocomposite materials [5, 6]. These materials induce high surface area, large pore volume, good dispersion and strong adsorptivity. A synergetic effect is expected in the coexistence of ZnO and a support which eventually transfer species from the support to ZnO or vice versa by the interface created between two phases. Hence, an increase in reaction rates is expected and ZnO/support composites have been postulated as suitable alternative photocatalysts in environmental applications.

The adsorption on the cenospheres and/or plerospheres of fly ashes makes use of them as ideal support materials. Synthesis, characterization, adsorption property and photoactivity of ZnO- or TiO_2-loaded fly ash composites are recently examined for anti-corrosion in coatings and also degradation of inorganic and organic pollutants to provide additional way to utilize the waste fly ash [7–13]. Photoactive ZnO nanoparticles supported on sepiolite are also reported for applications in decontamination of pollutants [14–19]. The immobilization of ZnO nanoparticles can be improved within the sepiolite framework due to the net negative charge of the sepiolite. Thus, charge separation efficiency and high adsorption capability can enhance the removal performance for the photodecomposition of organic pollutants.

This chapter is focused on a ternary photocatalyst; ZnO-supported fly ash-sepiolite (ZnO-FA-Sep) by taking advantage from the above-mentioned unique characteristics of fly ash and sepiolite. This composite can be considered as the first approximation of in situ formation of ZnO nanoparticles in presence of a clay (Sep) and an ash material (FA). With this objective, the as-prepared supported catalysts are characterized, structurally examined and evaluated via adsorption ability, kinetics and photocatalytic performance.

2. Preparation and characterization of the photocatalysts

ZnO catalysts were synthesized via a co-precipitation route [20]. In a typical preparation step, 0.5 M (or 0.25 or 0.125 M) $Zn(NO_3)_2{\cdot}6H_2O$ was added gradually to 0.5 M (or 0.25 or 0.125 M)

Na_2CO_3 solution under vigorous stirring for 2 h at room temperature. The resulting product was collected by centrifugation, washed with deionized water several times, dried at 100°C overnight and calcined at 500°C. Finally, the samples were designated as 0.125 M ZnO, 0.25 M ZnO and 0.5 M ZnO.

Raw Sep and FA used as supports in this study were obtained from Eskişehir-region of Anatolia and Soma (Turkey), respectively. They were characterized by X-ray diffraction and SEM-EDX analyses [18, 19]. Supported catalysts were prepared with the addition of FA-Sep dispersions (1:1 w/w, stirred about 12 h) into the above-mentioned ZnO solutions. The resulting product was collected by centrifugation, washed with deionized water for several times, dried at 100°C overnight and calcined at 500°C. Finally, the catalysts were designated as 0.125 M ZnO-FA-Sep, 0.25 M ZnO-FA-Sep and 0.5 M ZnO-FA-Sep.

X-ray diffraction patterns were recorded on an X-ray diffractometer (Rigaku-D/MAX-Ultima-diffractometer, 40 kV, and 40 mA) equipped with graphite monochromatized Cu-Kα1 radiation (λ = 1.54 Å) at a scan rate of 2° min^{-1}. Brunauer-Emmett-Teller (BET)-specific surface area and pores size were determined from the nitrogen adsorption apparatus (Quantachrome Nova 2200e) at 77 K. Prior to the measurements, the samples were pretreated in a vacuum at 473 K for 24 h. The morphology of the products was investigated by scanning electron microscopy (ESEM-FEG/EDAX Philips XL-30) running at an accelerating voltage of 20 kV. The elemental composition of the composites was determined by energy dispersive X-ray spectroscopy (EDS). X-ray photoelectron spectroscopy (XPS) data were recorded with a Thermo Scientific K-Alpha Photoelectron Spectrometer using Al Kα (12.5 kV) X-ray source. Calibration of the instrument was done via carbon peak [21, 22]. UV-visible (UV-vis) absorption spectra were recorded with a Shimadzu UV-2450. Diffuse reflectance measurements (DRS) were recorded by using an integrated sphere reflectance accessory with $BaSO_4$ reference.

The photodegradation of methyl orange (MO)-probe molecule in this study was carried out at room temperature in a homemade photocatalytic reactor under UV light (Philips TL 15 W/5BLB, λ = 365 nm, an incident photon flux of 4.7×10^{15} photons s^{-1}) [12, 18]. A typical reaction system included 0.2 g of a catalyst and 200 mL of a known concentration of MO. The suspension was stirred in the dark for 30 min. Thereafter, irradiation was started and UV-vis absorption spectra were recorded to monitor both degradation and decolorization processes. The decrease of the band at 274 nm indicated degradation of MO's aromatic moiety while the one at 464 nm was followed for the decolorization of MO solution. All experiments were performed at room temperature and at pH = 8 (3.27 mg L^{-1} MO in the presence of 0.5 M ZnO-FA-Sep) without concerning the degradation intermediates. Also, measurements were conducted at least twice and the average value was recorded. The degradation and decolorization rate percentages of MO were calculated by the following equation:

$$\text{Degradation (or decolorization)}\% = \frac{C_0 - C}{C_0} \times 100 \tag{1}$$

where C_0 was the initial concentration of MO and C was the concentration of MO after "t" minutes irradiation.

3. Structural characterization

3.1. XRD analysis

XRD patterns of Sep, FA, FA-Sep and the supported catalysts are shown at both low-angle ($2\theta < 10^{\circ}$) and high-angle ($20^{\circ} < 2\theta < 60^{\circ}$) ranges (**Figure 1A**). The characteristic d 1 1 0 reflection of the Sep appears at 7.24° (2θ) with a basal spacing of 12.19 Å. The pattern of FA includes amorphous aluminosilicate glass (as major) phase. Quartz (SiO_2) reflections of d 1 0 0, d 1 0 1 and d 1 1 0 are detected at 20.8, 26.6 and 36.6° (2θ), respectively. Minor constituents of calcite ($CaCO_3$) reflections of d 1 0 1 and d 1 0 4 are observed at 30° and 48° (2θ), respectively. Lime (CaO) reflection at 62° (2θ) of d 104 is also noticed. For the FA-Sep support, the intensities of d 1 1 0 reflection and all other high angle peaks decrease. Supported catalysts reveal d 1 0 0, d 0 0 2, d 1 0 1, d 1 0 2, d 1 1 0, d 1 0 3 and d 2 0 0 crystal planes of ZnO at 31.9, 34.6, 36.4, 47.7, 56.7, 63.1 and 66.6° (2θ), respectively (JCPDS file no. 36-1451). The peak position and basal spacing of the d 1 1 0 reflection do not change significantly for the FA-Sep and supported catalysts due to the non-expandable nature of the Sep. The ZnO signals are intensified with the increasing loading concentrations while the retained FA-Sep peaks decrease significantly.

Scherrer's equation is used to evaluate ZnO crystalline sizes (D_{ZnO}) in the supported catalysts. According to the d 1 0 1 reflection of ZnO, D_{ZnO} values are found as 16.1, 13.1, 9.85 and 11.7 nm for 0.25 M ZnO, 0.125 M ZnO-FA-Sep, 0.25 M ZnO-FA-Sep and 0.5 M ZnO-FA-Sep, respectively (**Table 1**). Hence, formation of ZnO aggregates is inhibited in the presence of the FA-Sep support. The reduction in the ZnO crystalline sizes in the form of supported catalysts and the decrements in the FA-Sep reflections may suggest distribution of ZnO nanoparticles over the surface and bulk.

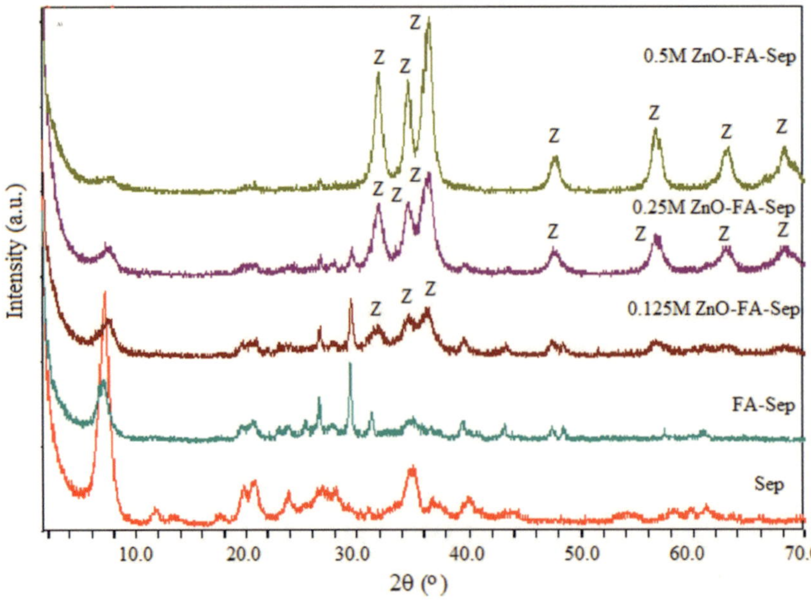

Figure 1A. XRD patterns of Sep, FA-Sep and supported catalysts (Z: ZnO).

Catalysts	D_{ZnO} (nm)[a]	BET ($m^2 g^{-1}$)[b]	V_{pore} ($cm^3 g^{-1}$)[c]	R_{pore} (Å)[d]
0.25 M ZnO	16.08	7.58	0.012	17.7
FA	–	1.80	0.003	14.9
Sep	–	104.5	0.140	14.9
FA-Sep	–	46.4	0.006	15.6
0.125 M ZnO-FA-Sep	13.11	61.7	0.111	14.9
0.25 M ZnO-FA-Sep	9.85	62.3	0.121	15.4
0.5 M ZnO-FA-Sep	11.66	50.2	0.097	14.8

[a]Calculated from the (1 0 1) diffraction peak of ZnO using the Scherrer equation.
[b]Determined from nitrogen adsorption–desorption isotherms using BET equation.
[c]Determined from cumulative adsorption pore volume using BJH method.
[d]Determined from adsorption pore size using BJH method.

Table 1. Crystalline sizes (D_{ZnO}), surface areas (BET), total pore volumes (V_{pore}) and pore radius (R_{pore}) of 0.25 M ZnO, supports and supported catalysts.

3.2. SEM (EDX) analysis

Figure 1B shows SEM images of Sep, FA, FA-Sep, 0.25 M ZnO-FA-Sep and elemental mapping images of the supported catalyst. The Sep exhibits a wavy pattern among the layers. The FA displays spherical (with air holes) and non-shaped particles. For the FA-Sep, the structures of the Sep and FA are retained with Si, Ca and Al major, Mg and Fe minor constituents. The

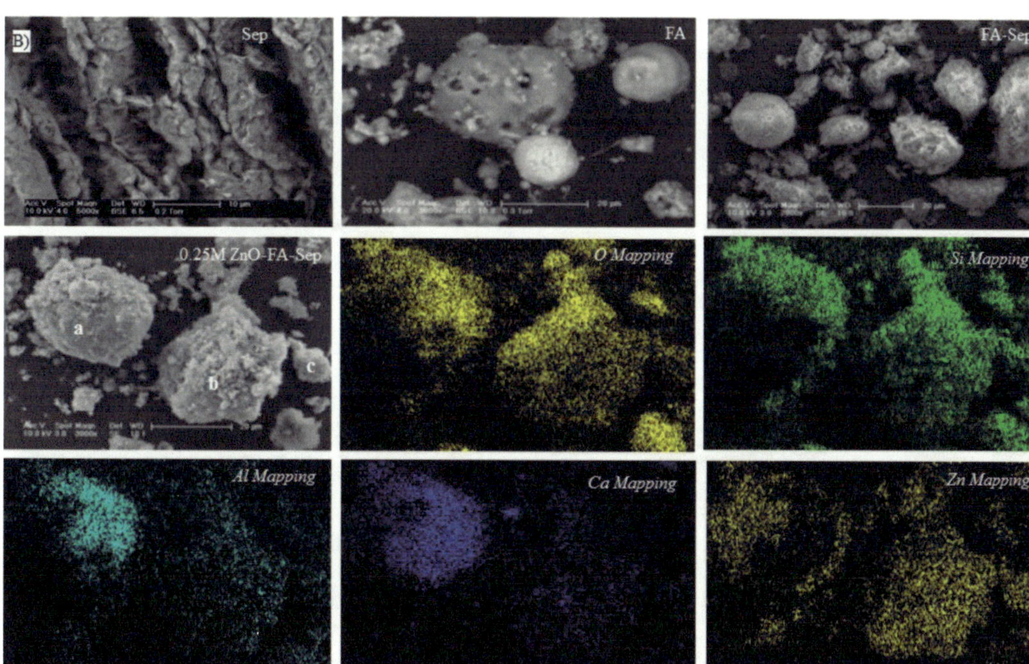

Figure 1B. SEM and mapping images of Sep, FA, FA-Sep and 0.25 M ZnO-FA-Sep.

appearance of 0.25 M ZnO-FA-Sep is different from the morphologies of Sep, FA and FA-Sep. The mixed structure of FA spheres and Sep layers suggests hosting of ZnO nanoparticles simultaneously on the spheres and/or non-shaped particles. This is proved by the EDX-spot analysis where Zn percentages are founds as 29 and 69% on the spheres 'a' and 'b' and 66% on the non-shaped particle 'c', respectively. Dominating Zn signals in the mapping images verify the dispersion of ZnO nanoparticles on the FA-Sep support.

3.3. Nitrogen adsorption-desorption isotherms

Nitrogen adsorption-desorption isotherms and pore size distributions are shown in **Figure 1C** and **D**, respectively. The non-porous FA and 0.25 M ZnO reveal Type II isotherms. For the Sep,

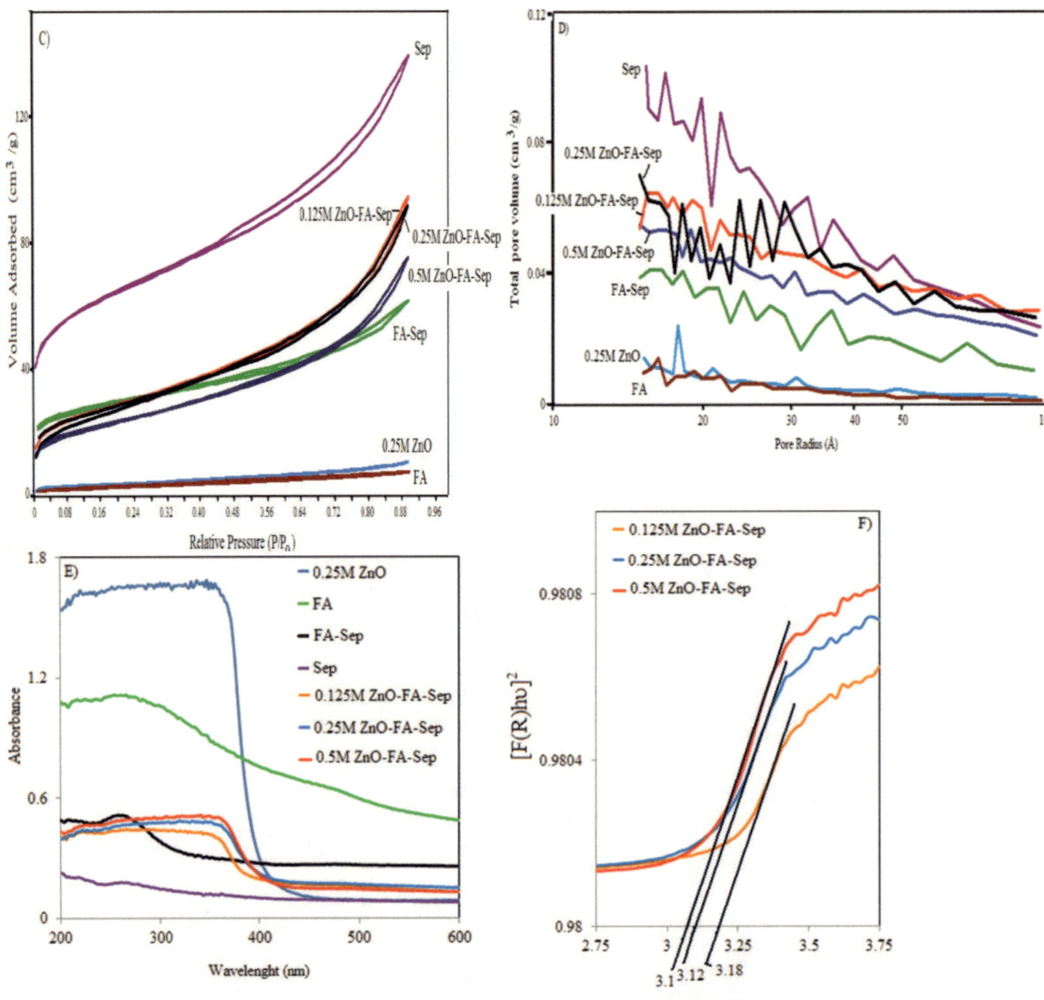

Figure 1C-F. (C) nitrogen adsorption/desorption isotherms of Sep, FA, FA-Sep, 0.25 M ZnO, supported catalysts, (D) pores size distribution plots of Sep, FA, FA-Sep, 0.25 M ZnO, supported catalysts, (E) diffuse reflectance spectra of Sep, FA,FA-Sep, 0.25 M ZnO, supported catalysts, (F) Kubelka-Munk transformed reflectance spectra of supported catalysts.

FA-Sep and supported catalysts, Type II isotherms are also detected with sites of different affinity to nitrogen and multilayer coverage at high P/P_0. Almost similar pore radius is detected for all catalysts and the supports (**Table 1**). The FA-Sep composite and supported catalysts exhibit lower surface areas and pore volumes in comparison to the Sep. This suggests an increment in the number of hosted FA spheres and/or ZnO nanoparticles. In the meantime, supported catalysts possess higher surface areas and pore volumes than the FA-Sep. Thus, enhanced adsorption capacities and degradation abilities are expected in the existence of the supported catalysts.

3.4. UV-vis DRS analysis

UV-vis absorption spectroscopies of the FA, Sep, FA-Sep and supported catalysts are presented in **Figure 1E**. Absorption profiles of the supports are similar with extensions in between 200 and 600 nm. 0.25 M ZnO shows its characteristic edge below 400 nm. All supported catalysts reveal this edge with a slight blue shift relative to 0.25 M ZnO and longer wavelength tails are not detected. The band gap energies for 0.125 M ZnO-FA-Sep, 0.25 M ZnO-FA-Sep and 0.5 M ZnO-FA-Sep are evaluated by linear extrapolation and taking the intercept on the *x*-axis as

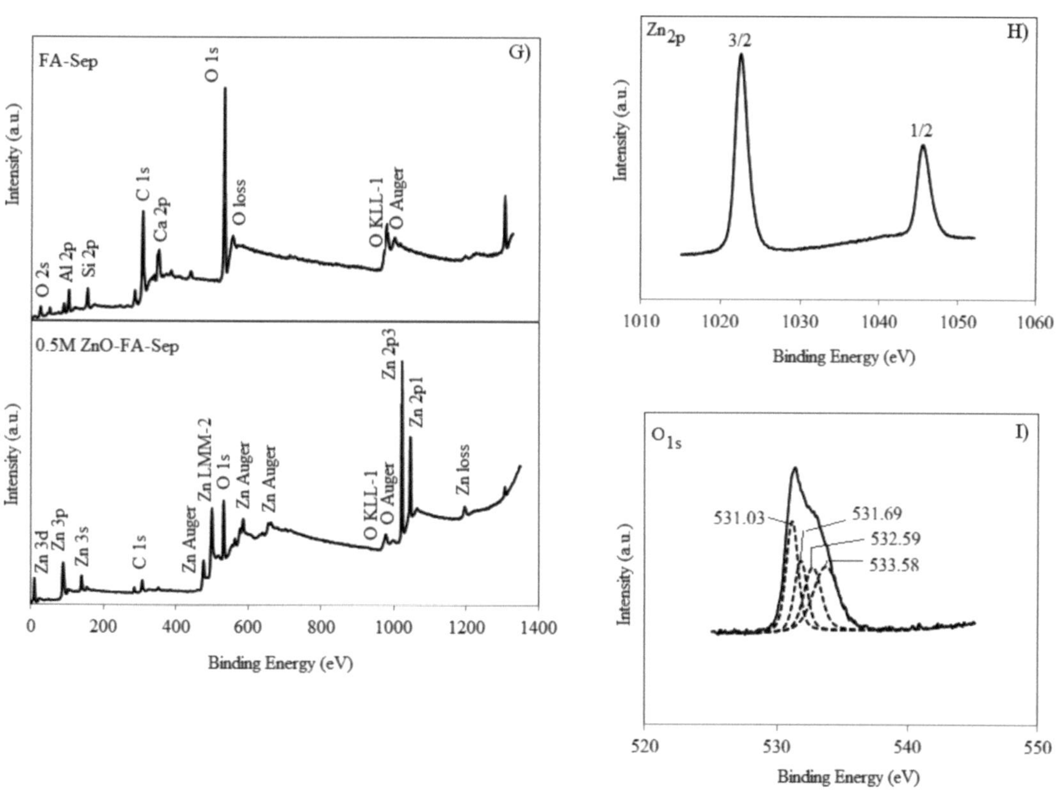

Figure 1G-I. (G) XPS survey analysis of FA-Sep and 0.5 M ZnO-FA-Sep, (H) Zn 2p XPS spectra of 0.5 M ZnO-FA-Sep, (I) O 1s XPS spectra of 0.5 M ZnO-FA-Sep.

3.10, 3.12 and 3.18 eV, respectively (**Figure 1F**). Thus, supported catalysts have suitable band gap energies for the degradation of MO under UV-A irradiation.

3.5. XPS analysis

XPS analysis is also performed to control the surface structure of the 0.5 M ZnO-FA-Sep. The survey scan reveals Zn peaks in addition to the Al (75.6 eV), Si (103 eV), Ca (347.07 eV), O (531.47 eV) and some Auger peaks of the FA-Sep support (**Figure 1G**). The doublet in Zn peaks corresponds to Zn 2p3/2 and 2p1/2 core levels (**Figure 1H**). Zn exists mainly in the form of Zn^{2+} oxidation state on the catalyst surface owing to the sharpness in Zn 2p3/2 peak [23]. The broad O 1 s signal of the 0.5 M ZnO-FA-Sep catalyst is deconvoluted by four subspectral parts (**Figure 1I**). CaO and Al_2O_3 (531.03 eV, 26.9% spectral area), oxygen in the form of $Zn(OH)_2$ (531.69 eV, 18.5% spectral area), SiO_2 (532.59 eV, 22.3% spectral area) and adsorbed water (533.58 eV, 32.2% spectral area) are labeled as the components of the supported catalyst.

4. Adsorption and photocatalytic applications

4.1. Control experiments

Prior to the evaluation of photoactivities of the supported catalysts in detail, two control experiments are designed. As a first step, photolysis of MO is tested under UV illumination and a negligible degradation is found (**Figure 2A**). Then, FA-Sep, 0.25 M ZnO-FA-Sep and 0.5 M ZnO-FA-Sep are examined without irradiation. In the presence of FA-Sep, the remaining MO percentages in the solution are detected as 84% for decolorization and 85% for degradation. However, supported catalysts reveal better adsorption capabilities in comparison to FA-Sep, which further enhanced with the ZnO loading concentration. Accordingly, 0.25 M ZnO-FA-Sep demonstrates 75% (for decolorization) and 73% (for degradation) MO staying in solution within 30 min and thereafter no significant change is noticed. Contrarily, 0.5 M ZnO-FA-Sep shows rapid decrements in both decolorization (7%) and degradation (5.3%) processes within 80 min.

MO exists in the anionic form in water since its natural pH (5.85) is higher than its pK_a (3.4) value [24]. The alkaline character of FA-Sep (pH = 10) solution, hence, induces repulsive forces among the MO molecules and FA-Sep. However, the weak interaction between positively charged species (Fe_2O_3, TiO_2 as minor phases) within the disordered FA-Sep structure and negatively charged MO moiety induce decrements in MO percentages. The attraction may proceed through the chromophoric (—N=N—) group and declines the absorption band at 464 nm. Simultaneously, weakening in the conjugation disrupts the benzenic rings and decreases the 274 nm band. The presence of ZnO nanoparticles in the supported matrixes is expected to increase the contact among the supported catalysts and MO molecules which eventually facilitate the degradation and decolorization processes.

In the second set, the photoactivities of the supported catalysts are controlled under irradiation (**Figure 2B**). The lowest MO remaining percentages with slower rates are obtained in the presence of 0.125 M ZnO-FA-Sep within 130 min (45% for decolorization and 47% for degradation). The increment in the ZnO concentration improves the performances of the catalysts.

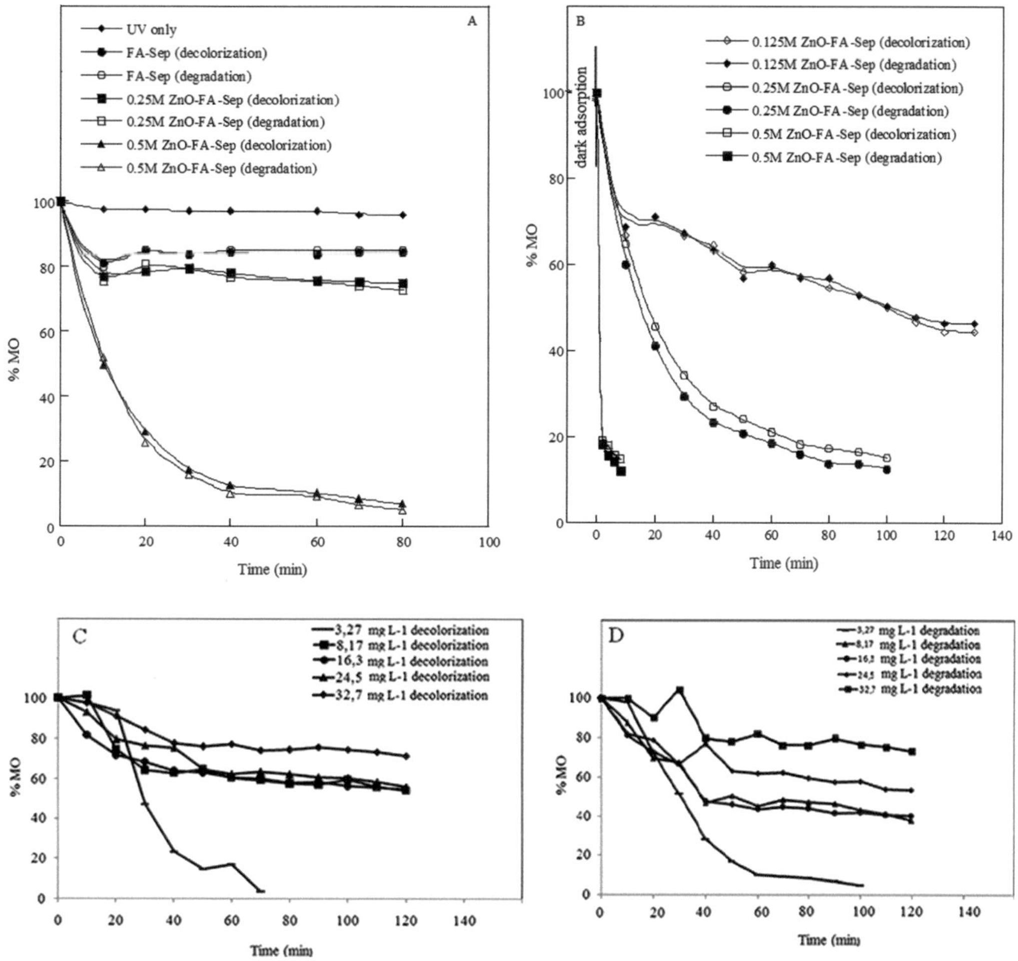

Figure 2. (A) Photolysis of MO and dark adsorption experiments of FA-Sep and supported catalysts, (B) photocatalytic activities of the supported catalysts, (C) MO remaining percentages for decolorization and (D) MO remaining percentages for degradation.

0.5 M ZnO-FA-Sep shows the best activity with the lowest MO remaining percentages (15% for decolorization and 12% for degradation) in less than 10 min. Rapid photodegradation of MO is facilitated, due to the synergistic effect of: (1) its adsorption on the mixed structures of FA and Sep, and (2) its photodegradation by the ZnO nanoparticles.

Dark adsorption capacity of 0.5 M ZnO-FA-Sep catalyst is also investigated by varying the initial MO concentrations from 3.27 to 32.7 mg L^{-1} (**Figure 2C** and **D**). It is evident that the amount of adsorbed MO at low initial concentrations is smaller than the corresponding amount at higher initial values, while MO removal percentages decrease significantly with increasing initial MO concentration. The highest MO remaining percentages are noticed with 32.7 mg L^{-1} MO concentration as 73.2 and 71.2% after 120 min for decolorization and degradation processes, respectively. The removal rate of MO is fast up to 30–40 min, and then gradually decreases with the increase in contact time due to the saturation on the catalyst surface. Contrarily, the lowest MO percentages (around 10% for decolorization and 16% for degradation) are obtained within 60 min for 3.27 mg L^{-1} initial MO concentration. Although surface active sites are more available, very fast adsorption rate and short extraction time are achieved for lower MO concentrations.

4.2. Adsorption equilibrium and kinetics

The isotherm analysis of the equilibrium data is examined by fitting the experimental data to Langmuir, Freundlich, Temkin and Dubinin and Radushkevich (D-R) isotherms to find the suitable model [25–31].

Langmuir isotherm: Langmuir isotherm assumes that the adsorption takes place at a specific homogeneous site within the adsorbent, all sites are equivalent, and there are no inteactions among the adsorbate molecules. The isotherm can be presented by the following equation [32]:

$$q_e = \frac{K_L a_L C_e}{1 + a_L C_e} \tag{2}$$

where K_L is the adsorption capacity and a_L is the energy of adsorption. Eq. (3) presents the linearized form of Langmuir isotherm. **Figure 3A** supplies data about the K_L (from the *y*-intercept) and a_L (from the slope) values

$$\frac{1}{q_e} = \frac{1}{K_L a_L C_e} + \frac{1}{K_L} \tag{3}$$

The theoretical maximum monolayer adsorption capacity of 0.5 M ZnO-FA-Sep, q_m (mg g^{-1}), is also calculated as 48.8 and 28.6 mg g^{-1} from the ratio of K_L to a_L for the decolorization and degradation processes, respectively (**Table 2**).

The separation factor R_L (dimensionless constant) indicates the favorable adsorption within 0–1 range [33] and depending on a_L and C_0 values as:

$$R_L = \frac{1}{1 + a_L C_0} \tag{4}$$

The R_L values in the MO concentration range of 3.27–32.7 mg L^{-1} are found to vary in between 0.58–0.12 and 0.51–0.094 for decolorization and degradation, respectively.

The correlation coefficients (0.917 for decolorization and 0.858 for degradation) are reasonable to suggest the applicability of the Langmuir model for the interpretation of the experimental data over the whole concentration range.

Freundlich isotherm: Freundlich isotherm predicts heterogeneous adsorption surface and active sites with different energy [34] and presented by the following equation [26] and **Figure 3B**.

$$q_e = K_f {C_e}^{1/n} \tag{5}$$

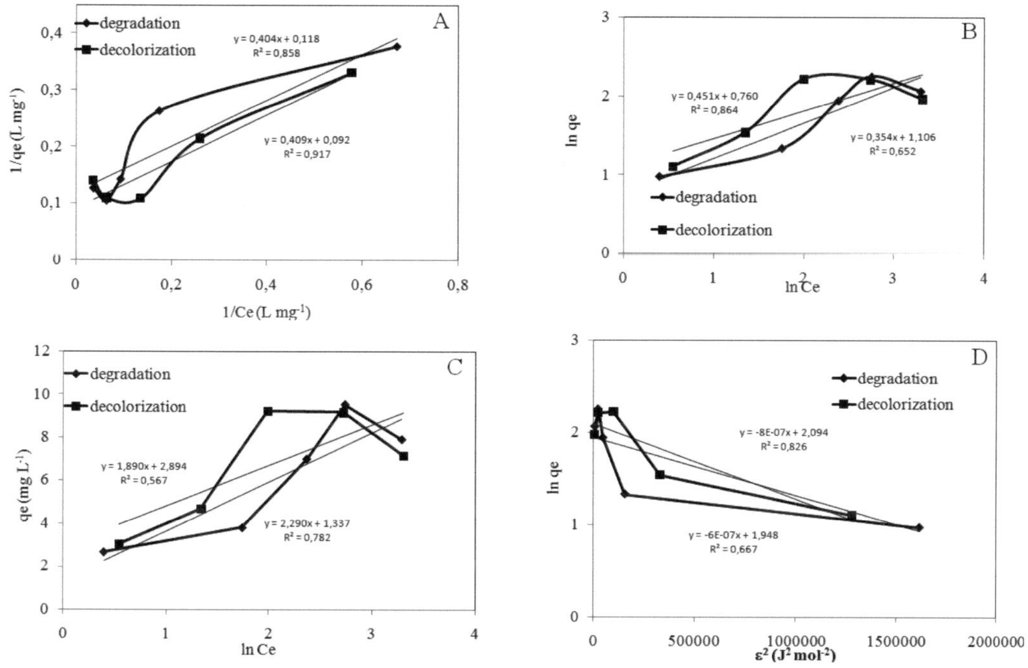

Figure 3. (A) Langmuir isotherm, (B) Freundlich isotherm, (C) Temkin isotherm, (D) Dubinin-Radushkevich isotherm.

	Decolorization	Degradation
Langmuir		
K_L	10.84	8.41
a_L	0.225	0.294
q_m	48.18	28.6
R^2	0.917	0.858
Freundlich		
KF	2.14	3.02
1/n	0.45	0.35
R^2	0.864	0.652
Temkin		
K_T	4.62	1.79
B_T	1.89	2.29
R^2	0.567	0.782
Dubinin-Radushkevich		
q_m	8.12	1.79
E	0.79	0.91
R^2	0.826	0.667

Table 2. Adsorption isotherm constants.

where q_e is equilibrium concentration (mg g^{-1}), C_e is equilibrium liquid phase concentration (mg L^{-1}) and K_f is adsorption capacity. Adsorption intensity is determined by "n". The values of 1/n are found as 0.451 (for decolorization) and 0.354 (for degradation) indicate high tendency of MO for the adsorption onto the supported catalyst. However, Freundlich model is not suitable to describe the relation between sorbed MO molecules and their equilibrium concentrations owing to the lower correlation coefficients (0.864 for decolorization and 0.652 for degradation).

Temkin isotherm: The heat of adsorption and the adsorption-binding energy relation are explored by the following Temkin equation [35]:

$$q_e = \frac{RT}{b} \ln\left(K_{Te}C_e\right) \tag{6}$$

Eq. (6) can be linearized as

$$q_e = B_T \ln K_T + B_T \ln C_e \tag{7}$$

where K_{Te} is equilibrium binding constant (L mg^{-1}), b is the heat of adsorption (J mol^{-1}), B_T is related to the heat of adsorption as being equal to RT/b, R is the gas constant and T is the temperature (K). The values of Temkin constants (K_{Te} = 4.62 and B_T = 1.89 for decolorization and K_{Te} = 1.79 and B_T = 2.29 for degradation) and correlation coefficients (R^2 = 0.567 for decolorization and R^2 = 0.782 for degradation) are found to be lower than the Langmuir values (**Table 2, Figure 3C**). Thus, Temkin model does not fit to the corresponding experimental data of MO adsorption on 0.5 M ZnO-FA-Sep.

Dubinin-Radushkevich (D-R) isotherm: Dubinin-Radushkevich model is based on heterogeneous surfaces, different sorption sites and steric hindrance effect among adsorbed molecules and bulk species [36]. The linear form of D-R isotherm is expressed by the following equation [31]:

$$\ln q_e = \ln q_m - \beta\varepsilon^2 \tag{8}$$

where q_e is the amount of adsorbate per unit weight of adsorbent (mg g^{-1}), q_m is the maximum adsorption capacity (mg g^{-1}), β is a coefficient for adsor ption mean free energy (mol^2 J^{-2}) and ε is the Polanyi potential ($\varepsilon = RT \ln(1 + 1/C_e)$).

The q_m values are found using the intercept of the plot as 8.12 and 7.02 mg g^{-1} for decolorization and degradation, respectively (**Table 2, Figure 3D**).

The adsorption mean free energy (E; kJ mol^{-1}) can be found from the slope as follows

$$E = \frac{1}{\sqrt{-2\beta}} \tag{9}$$

Its value gives information about adsorption mechanism whether it is physical or chemical. If it lies between 8 and 16 kJ mol^{-1}, the adsorption process takes place chemically and while E < 8 kJ mol^{-1}, the adsorption process proceeds physically [37, 38]. Since adsorption energies are calculated as 0.79 kJ mol^{-1} for decolorization and 0.91 kJ mol^{-1} for degradation processes, only physical interactions are probable among MO moiety and the supported catalyst. The lower correlation coefficients (R^2 = 0.826 for decolorization and R^2 = 0.667 for degradation) again represents the poorer fit of the experimental data.

The kinetics of MO adsorption onto 0.5 M ZnO-FA-Sep are studied in terms of pseudo-first order [39], pseudo-second order [40], Elovich [41–43] and intraparticle diffusion [44, 45] models.

Pseudo-first order model: Pseudo-first order model describes the adsorption rate based on the adsorption capacity [39].

The differential equation of the model is expressed as follows:

$$\frac{dq_t}{dt} = k_1(q_e - q_t) \tag{10}$$

where k_1 is the pseudo-first order rate constant (min), q_e is the adsorption capacity at equilibrium and q_t is the adsorption capacity at time t.

Integrating Eq. (10) for the boundary conditions of $q_t = 0$ at $t = 0$ and $q_t = q_t$ at $t = t$ gives

$$\ln(q_e - q_t) = \ln q_e - k_1 t \tag{11}$$

Accordingly, the values of $\ln(q_e - q_t)$ are linearly correlated with t by plot of $\ln(q_e - q_t)$ versus t. The rate constant (k_1) and q_e can be determined from the slope and intercept of the plot, respectively. The pseudo-first order equation fits well for the first 30 min data and then deviations are noticed. Although rate should be proportional to the first power of MO concentration, the linearity is lost for the higher initial MO concentrations. This may be attributed to the limiting effect of pore diffusion through the adsorption process. For the rapid adsorption of the initial stages, the first order rate constants, k_1, decrease with increments in MO concentrations for both decolorization and degradation processes (**Table 3**). It is also noticed that calculated q_e values agree well with the experimental data. The high correlation coefficients show the applicability of pseudo-first order kinetics to the adsorption of MO onto the supported catalyst.

Pseudo-second-order equation: The adsorption kinetics may also be described by the pseudo-second order model [40]. The differential equation is given as follows:

$$\frac{dq_t}{dt} = k_2(q_e - q_t)^2 \tag{12}$$

where k_2 is the second-order rate constant of adsorption. Integrating Eq. (12) for the boundary conditions of $q_t = 0$ at $t = 0$ and $q_t = q_t$ at $t = t$ gives

$$\frac{t}{q_t} = \left(\frac{1}{k_2 q^2{}_e}\right) + \frac{t}{q_e} \tag{13}$$

The linearity obtained in the plot of t/q_t versus t results in k_2 value (as the intercept) and q_e value as the equilibrium adsorption capacity (from the slope). Similar to the pseudo-first order kinetics, k_2 decreases as the concentration of initial MO increases for both processes (**Table 3**). Although high correlation coefficients are obtained, calculated q_e values are not closer to the experimental values. Moreover, the predicted chemical adsorption cannot be applicable since only physical interactions are suggested between MO molecules and the supported catalyst via D-R model.

Elovich equation: The Elovich equation is based on the adsorption capacity and expressed as follows [46–48]:

$$\frac{dq_t}{dt} = \alpha \exp\left(-\beta_{qt}\right) \tag{14}$$

where β is the initial adsorption rate (mg g^{-1} min^{-1}) and α is the desorption constant (g mg^{-1}). The linearized form of Eq. (14) with boundary conditions of $q_t = 0$ at $t = 0$ and $q_t = q_t$ at $t = t$ is as follows

$$q_t = (1/\beta)\ln(\alpha\,\beta) - (1/\beta)\ln t \tag{15}$$

Elovich parameters are given in **Table 3**. The plot of q_t versus ln t yields a linear relationship with a slope of 1/ β and an intercept of (1/β) ln (α β). The high correlation coefficients indicate suitability of the model for the evaluation of the adsorption process. The limited number of vacant-available sites on the supported catalyst may decrease the possibility of chemical adsorption process through increments in MO concentration. This eventually results in smaller β values in both decolorization and degradation processes. Simultaneously, desorptions from the surface may be enhanced via less strong physical attractions and increase α values.

Interparticle diffusion model: The adsorbate species are probably transported from the bulk of the solution into the solid phase through intraparticle diffusion/transport process. The model is expressed by the following equation:

$$q_t = K_{diff}t^{1/2} + C \tag{16}$$

where C (mg g^{-1}) is the intercept and K_{diff} is the intraparticle diffusion rate constant (in mg g^{-1} $min^{1/2}$). K_{diff} can be evaluated by the linear correlation of q_t versus $t^{1/2}$ (**Table 3**). As a general trend, K_{diff} values increase with increase of initial MO concentration but for the highest concentrations there seems to be a constancy. This may point out the existence of restricted diffusion for the external mass. Although C values supplies information about the thickness of the boundary layer, experimental data do not exhibit neither an increasing nor a decreasing trend. Thus, the adsorption mechanism cannot be explained by using this model owing to the complex structure of the supported catalyst.

Models	Parameters	Decolorization					Degradation				
		3.27 mg L^{-1}	**8.17 mg L^{-1}**	**16.3 mg L^{-1}**	**24.5 mg L^{-1}**	**32.7 mg L^{-1}**	**3.27 mg L^{-1}**	**8.17 mg L^{-1}**	**16.3 mg L^{-1}**	**24.5 mg L^{-1}**	**32.7 mg L^{-1}**
First-order kinetic model	k_1	0.035	0.032	0.026	0.025	0.0021	0.056	0.052	0.039	0.028	0.0039
	q_e	3.35	4.64	10.17	9.67	6.88	2.65	4.55	6.62	9.20	9.02
	R^2	0.897	0.934	0.949	0.971	0.912	0.984	0.727	0.985	0.924	0.906
Second-order kinetic model	k_2	3.83×10^{-3}	1.77×10^{-3}	1.62×10^{-3}	8.36×10^{-4}	7.47×10^{-4}	0.0508	7.65×10^{-3}	2.12×10^{-3}	1.38×10^{-3}	1.02×10^{-3}
	q_e	6.45	6.89	16.94	16.94	9.09	5.43	4.85	9.17	16.67	13.51
	R^2	0.697	0.941	0.846	0.856	0.852	0.916	0.987	0.998	0.903	0.898
Elowich	α	0.158	0.383	0.464	0.510	0.560	0.273	0.526	0.651	1.095	1.256
	β	0.683	0.621	0.248	0.229	0.219	0.633	0.534	0.442	0.273	0.271
	R^2	0.958	0.886	0.951	0.920	0.647	0.925	0.821	0.985	0.952	0.914
Intraparticle diffusion	K_{diff}	0.362	0.385	1.08	1.054	0.975	0.397	0.372	0.747	0.886	0.865
	C	0.12	1.45	0.29	0.76	1.67	0.153	0.931	0.978	2.14	1.513
	R^2	0.936	0.874	0.924	0.922	0.66	0.903	0.813	0.937	0.947	0.882
Experimental data	q_e(exp)	3.02	4.67	9.21	9.15	7.16	2.65	3.79	6.99	9.52	7.90

Table 3. Adsorption kinetics constants.

4.3. Photocatalytic degradation

4.3.1. Kinetics

The 0.5 M ZnO-FA-Sep is further used to analyze the rate of reactions in the initial concentration range of MO from 3.27 to 24.5 mg L^{-1}(**Figure 4A**). The data from the photocatalytic activities is analyzed for the following rate expression

$$\ln\frac{C_0}{C} = kt \tag{17}$$

where C_0 is taken as the equilibrium concentration of MO (mg L^{-1}) after dark adsorption. The linearity in the plot of $\ln(C_0/C)$ versus t results confirmed the validity of pseudo-first-order kinetics. The rate-constants (k, min^{-1}) are calculated from the slopes of the lines.

For the initial concentration of 3.27 mg L^{-1}, 88% degradation and 85% decolorization are achieved within 20 min whereas only 44% (degradation) and 43% (decolorization) MO removal are obtained even at 70 min for the concentration of 24.5 mg L^{-1}.Similarly, k values decrease from 0.122 to 0.036 min^{-1} (for degradation) and from 0.121 to 0.056 min^{-1} (for decolorization) as the initial concentration of MO increases from 3.27 to 24.5 mg L^{-1}(not shown). Since more MO molecules are expected to be adsorbed to the surface of 0.5 M ZnO-FA-Sep as the initial concentration of MO increases, the sorption of both OH^- and O_2 will be reduced. Thus, the number of both photogenerated holes and $^{\bullet}OH$ radicals is suppressed and resulted in lower photocatalytic efficiencies. Meanwhile, more photons are absorbed in bulk solutions by the high concentrations of MO. This creates the shortage of photons to activate the supported catalyst.

Additionally, Langmuir-Hinshelwood model is successfully applied to estimate the relationship between the photocatalytic degradation rate and the initial concentration of organic contaminants by the following rearranged form [49, 50].

$$\frac{1}{R} = \frac{1}{kKC_0} + \frac{1}{k} \tag{18}$$

where R is the rate of decolorization (or degradation), K is the adsorption coefficient of MO onto the 0.5 M ZnO-FA-Sep (L mg^{-1}), k is the reaction rate constant (mg L^{-1} min^{-1}). The linear correlation [0.997 (for degradation) and 0.998 (for decolorization)] in the plot of (1/R) against ($1/C_0$) proved the validity of the model for the as-prepared supported catalysts (**Figure 4B**). Accordingly, both adsorption of MO to 0.5 M ZnO-FA-Sep (K) and photocatalytic degradation of MO by 0.5 M ZnO-FA-Sep (k) are calculated as 0.066 L mg^{-1} and 0.200 mg L^{-1} min^{-1} (for degradation) and 0.177 L mg^{-1} and 0.154 mg L^{-1} min^{-1} (for decolorization).

4.3.2. Comparison with binary composites

The ternary-supported catalyst (0.5 M ZnO-FA-Sep) is also compared with the binary ones (0.5 M ZnO-Sep and 0.5 M ZnO-FA) and 0.5 M ZnO (**Figure 4C**). After 20 min dark adsorption, supported catalysts show lower MO percentages remaining in the solution than the 0.5 M ZnO (97% for decolorization and 91% for degradation). Meanwhile, 0.5 M ZnO-Sep displays the

http://dx.doi.org/10.5772/intechopen.70504

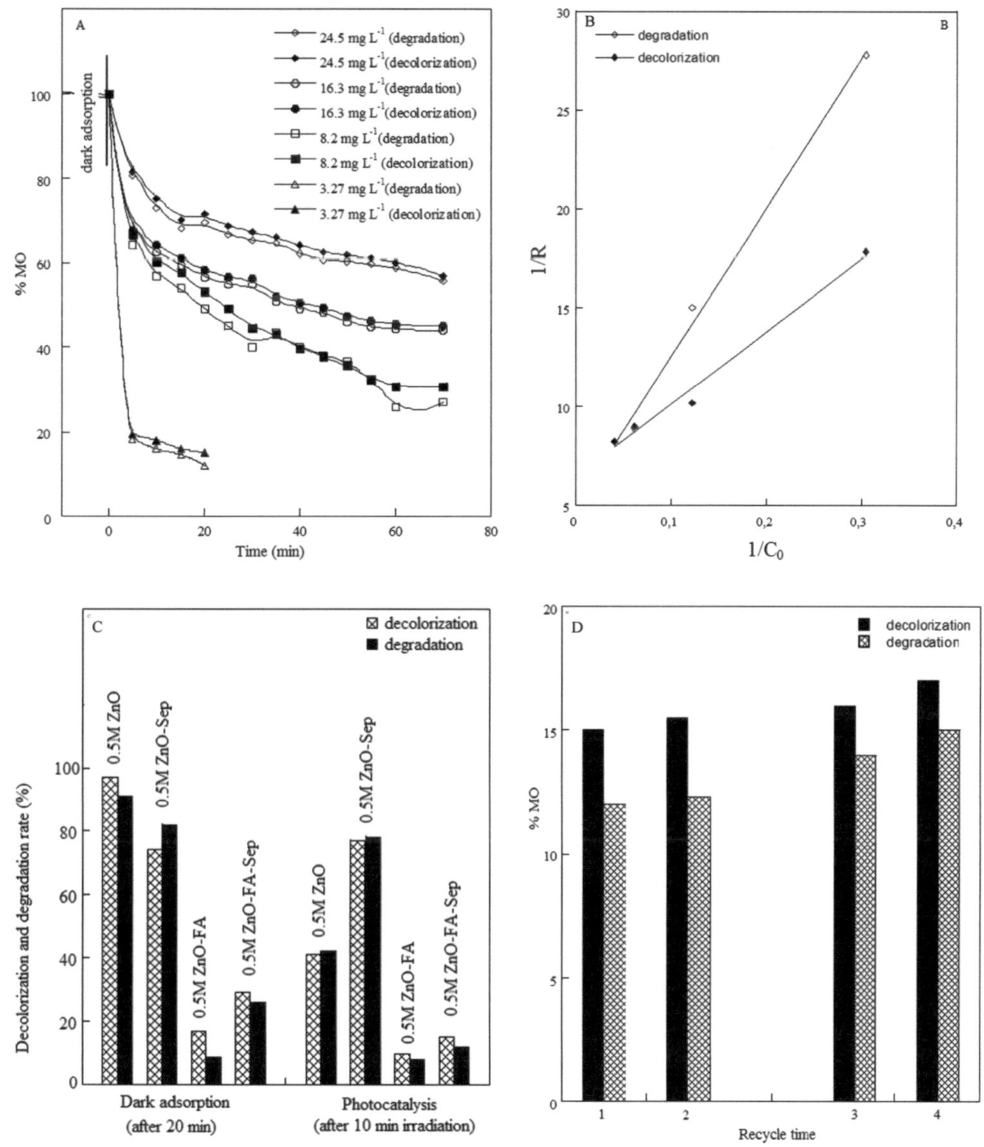

Figure 4. (A) Effect of initial MO concentration on the photoactivity of 0.5 M ZnO-FA-Sep, (B) Langmuir-Hinshelwood kinetic analysis in the presence of 0.5 M ZnO-FA-Sep, (C) photoactivity comparison between 0.5 M ZnO-FA-Sep, binary catalysts and 0.5 M ZnO, (D) reuse properties of 0.5 M ZnO-FA-Sep.

highest percentages (74% for decolorization and 82% for degradation) among the binary catalysts. However, the existence of FA spheres within the catalyst matrixes decreases the MO percentages. The lowest values (17% for decolorization and 9% for degradation) are noticed in the presence of 0.5 M ZnO-FA while relatively higher values (29% for decolorization and 26% for degradation) are obtained with 0.5 M ZnO-FA-Sep. Under irradiation, binary catalyst (0.5 M ZnO-Sep) does not work efficiently while the others demonstrate much lower percentages. Among which, 0.5 M ZnO-FA-Sep competes with the 0.5 M ZnO-FA within 10 min irradiation. The slight difference in the degradation and decolorization percentages may be due the smaller variations in the ZnO crystalline sizes, BET areas and pore volumes (D_{ZnO} = 8.12 nm, BET = 60 $m^2 g^{-1}$, V_{pore} = 0.122 $cm^3 g^{-1}$ for 0.5 M ZnO-FA and D_{ZnO} = 11.6 nm, BET = 50.2 $m^2 g^{-1}$, V_{pore} = 0.097 $cm^3 g^{-1}$ for 0.5 M ZnO) [12]. Although FA-Sep has a more complex structure than the FA alone, ZnO nanoparticles in the ternary system are located on the external surfaces of both FA and Sep showing no preference to any of these minerals as detected in the SEM images. Moreover, FA as well as Sep crystals do not agglomerate in the ternary composite owing to their morphological differences in shape and sizes. This creates uniqueness for 0.5 M ZnO-FA-Sep and enhances dispersion of ZnO nanoparticles.

4.3.3. Reusability

Furthermore, to investigate the photocatalytic stability of 0.5 M ZnO-FA-Sep, cyclic experiments are carried out under the same experimental conditions (**Figure 4D**). For each run, 0.5 M ZnO-FA-Sep is filtrated, washed and calcined at 500°C for 2 h. After four cycles, the percentage of MO remaining in solution is found to increase only approximately 2% (from 15 to 17% for decolorization process) and 3% (from 12 to 15% for degradation process). The slight increments in the percentages can be attributed to the catalyst loss during each collection and rinsing steps.

5. Conclusions

A ternary-supported catalyst has been prepared and characterized by XRD, BET, SEM (EDX), XPS and DRUV techniques. ZnO nanoparticles are found to be dispersed on both FA and Sep minerals. Supported catalysts exhibit higher surface areas and pore volumes than FA-Sep, FA and 0.25 M ZnO·D_{ZnO} sizes do not exhibit significant differences depending on the ZnO loading concentrations. EDX, mapping and XPS analysis evidence the existence of ZnO nanoparticles. Moreover, supported catalysts exhibit an absorption edge similar to 0.25 M ZnO with a slight blue shift. High dark adsorption capacities of the supported catalysts improve their photocatalytic activities, which further enhanced with the ZnO loading concentration. The existence of both FA and Sep with different shapes and sizes decrease their agglomeration and expose more of their surfaces for the adsorption of ZnO nanoparticles.

Future studies will focus on the development of new composites and their applications in environmental issues.

Acknowledgements

This study was supported by Boğaziçi University Research Foundation (Project No. 17B05P4/13021).

Author details

Ayşe Neren Ökte

Address all correspondence to: okteayse@boun.edu.tr

Department of Chemistry, Boğaziçi University, Istanbul, Turkey

References

[1] Yadava KP, Tyagi BS, Singh VN. Effect of temperature on the removal of lead (II) by adsorption on China clay and wollastonite. Journal of Chemical Technology and Biotechnology. 1991;**51**:47-60

[2] Afify AS, Hassan M, Piumetti M, Peter I, Bonelli B, Tulliani JM. Elaboration and characterization of modified sepiolites and their humidity sensing features for environmental monitoring. Applied Clay Science. 2015;**115**:165-173

[3] Ma Y, Zhang GK. Sepiolite nanofiber-supported platinum nanoparticle catalysts toward the catalytic oxidation of formaldehyde at ambient temperature: Efficient and stable performance and mechanism. Chemical Engineering Journal. 2016;**288**:70-78

[4] Li FF, Dai YZ, Gong M, Yu TP, Chen XJ. Synthesis, characterization of magnetic-sepiolite supported with TiO_2 and the photocatalytic performance over Cr(VI) and 2,4-dichlorophenol co-existed wastewater. Journal of Alloys and Compounds. 2015;**638**:435-442

[5] Wang B, Zhang GX, Leng X, Sun ZM, Zheng SL. Characterization and improved solar light activity of vanadium doped TiO_2/diatomite hybrid catalysts. Journal of Hazardous Materials. 2015;**285**:212-220

[6] Padmanabhan SK, Pal S, Haq EU, Licciulli A. Nanocrystalline TiO_2-diatomite composite catalysts: Effect of crystallization on the photocatalytic degradation of rhodamine B. Applied Catalysis A: General. 2014;**485**:157-162

[7] Shi JW, Chen SH, Wang SM, Ye ZL, Wu P, Xu B. Favorable recycling photocatalyst TiO_2/CFA: Effects of calcination temperature on the structural property and photocatalytic activity. Journal of Molecular Catalysis A: Chemical. 2010;**330**:41-48

[8] Wang B, Li C, Pang J, Qing X, Zhai J, Li Q. Novel polypyrrole-sensitized hollow TiO_2/fly ash cenospheres: Synthesis, characterization, and photocatalytic ability under visible light. Applied Surface Science. 2012;**258**:9989-9996

[9] Visa M, Duta A. Methyl-orange and cadmium simultaneous removal using fly ash and photo-Fenton systems. Journal of Hazardous Materials. 2013;**244–245**:773-779

[10] Visa M, Duca A. TiO_2/fly ash novel substrate for simultaneous removal of heavy metals and surfactants. Journal of Chemical Engineering. 2013;**223**:860-868

[11] Yeole K, Kadam P, Mhaske S. Synthesis and characterization of fly ash-zinc oxide nanocomposite. Journal of Materials Research and Technology. 2014;**3**:186-190

[12] Ökte AN, Karamanis D. A novel photoresponsive ZnO-flyash nanocomposite for environmental and energy applications. Applied Catalysis B: Environmental. 2013;**142–143**: 538-552

[13] Ökte AN, Karamanis D, Tuncel D. Dual functionality of TiO_2-flyash nanocomposites: Water vapor adsorption and photocatalysis. Catalysis Today. 2014;**230**:205-213

[14] Xu WG, Liu SF, Lu SX, Kang SY, Zhou Y, Zhang HF. Photocatalytic degradation in aqueous solution using quantum-sized ZnO particles supported on sepiolite. Journal of Colloid and Interface Science. 2010;**351**:210-216

[15] Bautista FM, Campelo JM, Luna D, Luque J, Marinas JM. Vanadium oxides supported on TiO_2-Sepiolite and Sepiolite: Preparation, structural and acid characterization and catalytic behaviour in selective oxidation of toluene. Applied Catalysis A: General. 2007; **325**:336-344

[16] Arques A, Amat AM, Santos-Juanes L, Vercher RF, Mar´ın ML, Miranda MA. Sepiolites as supporting material for organic sensitisers employed in heterogeneous solar photocatalysis. Journal of Molecular Catalysis A: Chemical. 2007;**271**:221-226

[17] Ökte AN. Advanced Catalytic Materials, Photocatalysis and Other Current Trends. Croatia: INTECH; 2016

[18] Ökte AN, Sayinsöz E. Characterization and photocatalytic activity of TiO_2 supported sepiolite catalysts. Separation and Purification Technology. 2008;**62**:535-543

[19] Karamanis D, Ökte AN, Vardoulakis E, Vaimakis T. Water vapor adsorption and photocatalytic pollutant degradation with TiO_2-sepiolite nanocomposites. Applied Clay Science. 2011;**53**:181-187

[20] Anandan S, Vinu R, Sheeja Lovely KLP, Gokulakrishnan N, Srinivasu P, Mori T, Murugesan V, Sivamurugan V, Ariga K. Photocatalytic activity of La-doped ZnO for the degradation of monocrotophos in aqueous suspension. Journal of Molecular Catalysis A: Chemical. 2007;**266**:149-157

[21] Lupan O, Chow L, Chai G, Roldan Cuenya B, Naitabdi A, Schulte A, Heinrich H. Nanofabrication and characterization of ZnO nanorod arrays and branched microrods by aqueous solution route and rapid thermal processing. Materials Science and Engineering B. 2007;**145**:57-66

[22] Lupan O, Emelchenko GA, Ursaki VV, Chai G, Redkin AN, Gruzintsev AN, Tiginyanu IM, Chow L, Ono LK, Roldan Cuenya B, Heinrich H, Yakimov EE. Synthesis and characterization of ZnO nanowires for nanosensor applications. Materials Research Bulletin. 2010;**45**:1026-1032

[23] Jing L, Xu Z, Sun X, Shang J, Cai W. The surface properties and photocatalytic activities of ZnO ultrafine particles. Applied Surface Science. 2011;**180**:308-314

[24] Guo P, Wang X, Guo H. TiO_2/Na-HZSM-5 nano-composite photocatalyst: Reversible adsorption by acid sites promotes photocatalytic decomposition of methyl orange. Applied Catalysis B: Environmental. 2009;**90**:677-687

[25] Langmuir I. The constitution and fundamental properties of solids and liquids. Part I. Solids. Journal of the American Chemical Society. 1916;**38**:2221-2295

[26] Freundlich HMF. Over the adsorption in solution. The Journal of Physical Chemistry. 1906;**57**:385-470

[27] Ghaedi M, Khajesharifi H, Hemmati Yadkuri A, Roosta M, Sahraei R, Daneshfar A. Cadmium hydroxide nanowire loaded on activated carbon as efficient adsorbent for removal of Bromocresol Green. Spectrochimica Acta Part A. 2012;**86**:62-68

[28] Zhang H, Tang Y, Liu X, Ke Z, Su X, Cai D, Wang X, Liu Y, Huang Q, Yu Z. Improved adsorptive capacity of pine wood decayed by fungi Poria cocos for removal of malachite green from aqueous solutions. Desalination. 2011;**274**:97-104

[29] Temkin MJ, Pyzhev V. Recent modifications to Langmuir isotherms. Acta Physiochimica URSS. 1940;**12**:217-222

[30] Uluozlu OD, Sari A, Tuzen M, Soylak M. Biosorption of Pb (II) and Cr (III) from aqueous solution by lichen (*Parmelina tiliaceae*) biomass. Bioresource Technology. 2008;**99**:2972-2980

[31] Sari A, Tuzen M. Kinetic and equilibrium studies of biosorption of Pb(II) and Cd(II) from aqueous solution by macrofungus (*Amanita rubescens*) biomass. Journal of Hazardous Materials. 2009;**164**:1004-1011

[32] Langmuir I. The adsorption of gases on plane surfaces of glass, mica and platinum. Journal of the American Chemical Society. 1918;**40**:1361-1403

[33] McKay G, Blair HS, Garden JR. Adsorption of dyes on chitin. 1. Equilibrium studies. Journal of Applied Polymer Science. 1982;**27**:3043-3057

[34] Gunay A, Arslankaya E, Tosun I. Lead removal from aqueous solution by natural and pretreated clinoptilolite: Adsorption equilibrium and kinetics. Journal of Hazardous Materials. 2007;**146**:362-337

[35] Temkin MJ, Pyzhev V. Recent modifications to Langmuir isotherms. Acta Physiochimica URSS. 1940;**12**:217-222

[36] Deliyanni EA, Nalbandian L, Matis KA. Adsorptive removal of arsenites by a nanocrystalline hybrid surfactant–akaganeite sorbent. Journal of Colloid and Interface Science. 2006;**302**:458-466

[37] Dubinin MM, Radushkevich LV. Equation of the characteristic curve of activated charcoal. Chemisches Zentralblatt. 1947;**1**:875

[38] Dubinin MM. The potential theory of adsorption of gases and vapors for adsorbents with energetically non-uniform surface. Chemical Reviews. 1960;**60**:235-266

[39] Dubinin MM. Modern state of the theory of volume filling of micropore adsorbents during adsorption of gases and steams on carbon adsorbents. Zhurnal Fizicheskoi Khimii. 1965;**39**:1305-1317

[40] Radushkevich LV. Potential theory of sorption and structure of carbons. Zhurnal Fizicheskoi Khimii. 1949;**23**:1410-1420

[41] Kundu S, Gupta AK. Investigation on the adsorption efficiency of iron oxide coated cement (IOCC) towards As(V)—kinetics, equilibrium and thermodynamic studies. Colloids and Surfaces A: Physicochemical and Engineering Aspects. 2006;**273**:121-128

[42] Lagergren S. Zur theorie der sogenannten adsorption geloster stoffe kungligasvenska vetenskapsakademiens. Handlingar. 1898;**24**:1-39

[43] Ho YS, McKay G, Wase DAJ, Foster CF. Study of the sorption of divalent metal ions on to peat. Adsorption Science & Technology. 2000;**18**:639-650

[44] Chien SH, Clayton WR. Application of Elovich equation to the kinetics of phosphate release and sorption on soils. Soil Science Society of America Journal. 1980;**44**:265-268

[45] Sparks DL. Kinetics of Reaction in Pure and Mixed Systems in Soil Physical Chemistry. Boca Raton: CRC Press; 1986

[46] Chang Y, Lye ML, Zeng HC. Large-scale synthesis of high-quality ultralong copper nanowires. Langmuir. 2005;**21**:3746-3748

[47] Sahraei R, Motedayen Aval G, Goudarzi A. Compositional, structural, and optical study of nanocrystalline ZnS thin films prepared by a new chemical bath deposition route. Journal of Alloys and Compounds. 2008;**466**:488-492

[48] Ghaedi M, Ansari A, Habibi MH, Asghari AR. Removal of malachite green from aqueous solution by zinc oxide nanoparticle loaded on activated carbon: Kinetics and isotherm study. Journal of Industrial and Engineering Chemistry. 2014;**20**:17-28

[49] Yang L, Yu LE, Ray MB. Degradation of paracetamol in aqueous solutions by TiO_2 photocatalysis. Water Research. 2008;**42**:3480-3488

[50] Yang H, Li G, An T, Gao Y, Fu J. Photocatalytic degradation kinetics and mechanism of environmental pharmaceuticals in aqueous suspension of TiO_2: A case of sulfa drugs. Catalysis Today. 2010;**153**:200-207

Section 3

Kinetics Techniques

Chapter 10

Kinetics of Heterogeneous Self-Propagating High-Temperature Reactions

Christopher E. Shuck and Alexander S. Mukasyan

Additional information is available at the end of the chapter

http://dx.doi.org/10.5772/intechopen.70560

Abstract

In this chapter, we present an overview of experimental techniques utilized and kinetic data collected for exothermic self-sustained noncatalytic heterogeneous reactions. The data focuses on five primary experimental techniques: electrothermal explosion, differential thermal analysis, electrothermography, combustion velocity/temperature analyses, and several advanced in situ diagnostics, including time-resolved X-ray diffraction.

Keywords: self-propagating high-temperature synthesis (SHS), electrothermal explosion, electrothermography, combustion synthesis, mechanical activation, high-energy ball milling, intermetallics, thermites

1. Introduction

Self-propagating high-temperature synthesis (SHS) is a technological approach for fabrication of materials, which involves self-sustained noncatalytic reactions [1–3]. Currently, there exist three major types of SHS systems: (i) gasless; (ii) with gasification of initially solid precursors; (iii) gas-solid. Within gasless systems, there are three major types of chemical reactions that occur: metal-metal, producing intermetallic (e.g., NiAl, NiTi), metal-nonmetal, leading to synthesis of borides, carbides and silicides (e.g., TiB_2, TaC, $MoSi_2$); nonmetal-nonmetal, producing ceramics (e.g., B_4C, SiC). In the gasification reactions, one of the precursors is volatile, including such elements as S, Se, P, As, Sb. Finally, the gas-solid systems include reactions between metals or nonmetals with different gases, such as nitrogen, oxygen, hydrogen, CO, and CO_2 leading to formation of nitrides, oxides, hydrides and etc. These lists are by no means exhaustive, as different research groups are continually exploring the limits of SHS reactions with different systems, reactants, and conditions. In order to obtain materials with desired microstructures, and thus properties, one has to precisely control the synthesis conditions during SHS. These conditions are primarily defined by the kinetics of the chemical reactions taking place in the combustion wave.

In order to study and understand the kinetics of SHS reactions, it is important to examine the fundamentals of kinetics and how they relate to SHS itself. Let us start with general definitions. Thus, assuming that the concentrations of any initial reagent, c_i, and the product are uniformly distributed throughout the entire volume (homogeneous or quasi homogeneous cases), the chemical reaction rate can be expressed by the following equation:

$$W_i = \frac{dC_i}{dt} \text{ or } W_i = \frac{d\eta_i}{dt} \tag{1}$$

where W_i is the reaction rate for the ith reagent or product, $\eta_i = (c^0{}_i - c_i)/c_0$ is the degree of conversion for the ith reagent, $c^0{}_i$ is the initial concentration of the ith reagent, and t is time.

In the nineteenth century, C.M. Guldberg with P. Waage and N.N. Beketov independently formulated the law of mass action. This essentially states that the chemical reaction rates at a given point are proportional to the concentration (mass) of the reactants raised to a proportional exponent. Thus, for an elementary chemical reaction between two reagents A and B of the following form:

$$v_1 A + v_2 B \rightarrow C, \tag{2}$$

where v_1, v_2 are stoichiometric coefficients.

For this reaction, the law of mass action can be written in the form of a kinetic equation:

$$W = k c_A^{v_1} c_B^{v_2}, \tag{3}$$

where k is the reaction rate constant.

Along with the reactant concentration, the temperature affects the rate of the chemical reaction in a noncatalytic homogeneous reaction. However, the mechanisms of these processes are often unknown or too complicated. This is because the reactions occur in multiple steps, each of which has unique reaction rates. In order to describe the chemical kinetics, a single-step approximation is typically used. This states that the rate of the processes in the condensed state is generally a function of the temperature and degree of conversion:

$$\frac{d\eta}{dt} = F(T, \eta) \tag{4}$$

The single-step approximation employs the assumption that the function in Eq. (4) can be expressed as a product of two separable functions that are independent of each other; the first, $K(T)$, depends solely on the temperature, T, and the second, $\Phi(\eta)$, depends solely on the degree of conversion, η:

$$\frac{d\eta}{dt} = K(T)\Phi(\eta) \tag{5}$$

The temperature function $K(T)$ is generally considered to be the rate constant, while the conversion function $\Phi(\eta)$ is generally considered represent the process mechanism. It is assumed that the reaction mechanism is solely dependent on the conversion, and not the temperature. Eq. (3) resembles a single-step kinetic equation, even though it represents the kinetics of a complex condensed-phase process. The single-step kinetic approximation results in the substitution of a generally complex set of kinetic equations with the sole single-step kinetic equation. Eq. (5) represents a mathematical formulation of the single-step kinetic approximation. With few exceptions, the temperature function is exclusively expressed by the Arrhenius equation:

$$K(T) = A\exp\left(-E_a/RT\right) \tag{6}$$

where A and E_a are considered to be the pre-exponential factor and the activation energy, respectively, T is the absolute temperature, and R is the gas constant.

As our knowledge about the atomic and molecular structure of matter increased, coupled with the development of quantum mechanics, new directions in chemical kinetics have emerged. These directions are typically related to the interactions of individual atoms and molecules, which are more fundamental studies. The set of elementary events is called the reaction mechanism. Fundamental studies on the reaction mechanisms allow us to formulate physical explanations to the kinetic parameters (A, E_a, etc.), which were originally introduced as empirical constants. For example, the activation energy E_a is an energy barrier that must be overcome by molecules in the reaction mixture to reach an interatomic distance where they can from a chemical bond. From Eq. (5), it is clear that, if the concentration of substances or the temperature in the given system varies from point to point. Thus, it is impossible to introduce a common reaction rate for the entire system. In order to get closer to these ideal conditions, in classical kinetic experiments we must continuously mix the reagents and maintain a constant temperature by use of a thermostat.

In the case of heterogeneous reactions involving a condensed phase, where the reactants are not mixed on the molecular level, there is an additional parameter, which controls the rate of interaction, i.e., the contact surface area (S) between the reagents [2]. In this case, the rate of the chemical reactions can be represented as follows:

$$\frac{d\eta}{dt} = A \cdot S \cdot \Phi(\eta)\exp\left(-E_a/RT\right) \tag{7}$$

The presence of condensed phases complicates the reaction; this phase requires that transport plays a role in the reaction. Thus, in general, the kinetics of such reactions are determined both by the intrinsic rate of the chemical reaction and by mass transport (e.g., diffusion). The transport phenomena are essential for replenishing the reactants that were consumed in the reaction zone [4]. Describing the reaction rate is further complicated when the temperature of the reacting environment is changing with time. In this case, along with the processes of mass transfer and chemical reactions, it is necessary to consider the specifics of heat transfer mechanisms [5]. Typically, the activation parameters are obtained after experiments considering the dependences of time vs. temperature (for isothermal measurements), temperature vs. heating rate (for integral and incremental methods with linear heating rates), or from reaction rate vs. temperature. Considering the above limitations and complications, only the effective or apparent activation energy can truly be considered, as it includes both the intrinsic kinetics as well as processes of heat and mass transport.

2. Techniques for studying SHS kinetics

The task of accurately determining kinetics becomes even more complicated when accounting for the extremely high temperatures of SHS processes (>1800 K) and rapid heating rates (10^3–10^5 K/s). Such parameters are essentially impossible to achieve using conventional approaches for measurement of kinetics parameters. While standard nonisothermal TGA/DTA-based approaches [6] are still used to evaluate the kinetics of SHS reactions, several unique methods such as electrothermal explosion (ETE) [7] and electrothermography (ET) [8] were specifically developed to fit the experimental conditions of SHS reactions. Moreover, recently a variety of advanced in situ diagnostics, including time-resolved X-ray diffraction (TRXRD) [9], high-speed X-ray phase-contrast imaging [10], and high speed transmission electron microscopy (HSTEM) [11] were modified to obtain the kinetics of phase transformations during SHS reactions.

3. Electrothermal explosion

The ETE method was developed in 1977 to study the rapid, high-temperature kinetics that occur in SHS systems [7]. It relies on rapid, uniform preheating of the sample until adiabatic thermal explosion occurs. A representation of a typical ETE setup is shown in **Figure 1**. Briefly, the sample is clamped between two metallic electrodes with sufficient clamping pressure to ensure adequate contact. The power is then initiated, leading to preheating of the sample until a set T_{off} point. After initiation, the resulting time-temperature profiles are simultaneously collected across a number of high-speed photodiodes. In the commonly used ETA-100 system (Aloft, Inc., Berkley, CA), there are 16 photodiodes present, with 1 mm in between them; this corresponds to 0.5 mm spatial resolution. The photodiodes have a temporal resolution of 10^{-5} s and are accurate within 900–3000 K. Once the sample is heated to the selected T_{off} point, the equipment heating is halted, with the consequent rate of self-heating determined solely by the chemical reaction rate. Due to the experimental conditions, i.e., the rapid initial preheating, the reaction occurs in the adiabatic mode. Once thermal ignition occurs, analysis of the time-temperature profile enables extraction of the kinetic parameters (see details in [12]). This technique can be used to study the kinetics at temperatures much higher than can be achieved in other experiments. However, it is often limited in the systems that can be studied due to the stringent heating conditions caused by Joule preheating.

ETE has been used for different gasless SHS systems. These studies have provided valuable kinetic data in extremely high-temperature ranges that are essentially inaccessible by other

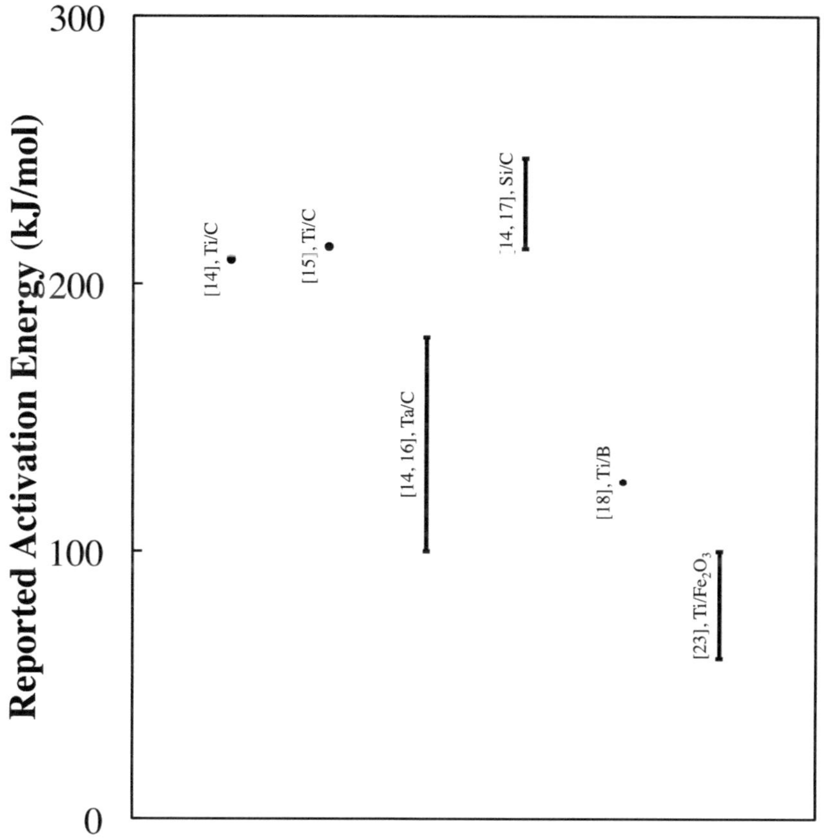

Figure 1. Summary of data collected using the electrothermal explosion technique.

methods. For carbides, the Ti/C system has been studied [13–15], with additional studies in the Ta/C [14, 16] and Si/C [14, 17] mixtures. The Ti/B system was also investigated [18]. The majority of ETE work being focused on the Ni/Al system [12, 19–21]. The Ti/Fe_2O_3 system is the only thermite system investigated by ETE [22]. In addition to the experimental studies, a number of theoretical models have been developed to better understand the ETE process [23–26]. **Figure 1** shows that ranges of reported activation energies for the above mentioned gasless systems, including both intermetallics and thermites obtained by ETE. It can be seen that the results are reproducible, confirming the reliability of the ETE approach. Additionally, the technique allows for a more complete understanding of systems that have multiple steps that rapidly occur in the high-temperature regime, which gives insight into the combustion process [2].

4. Differential thermal analysis/differential scanning calorimetry

This technique has been extensively used in many fields, including: polymer science, biochemistry, and materials science. In order to utilize these methods, a sample is heated at a constant rate until the maximum set temperature is reached. Throughout the experiment, the heat release characteristics of the sample are measured against an inert reference standard.

For typical systems, the points of differing heat release characteristics can be due to phase transitions, crystallization, or reactions; however, this section will only focus on SHS-reaction kinetics. In order to determine the reaction kinetics, the experiment is conducted multiple times with different heating rates. The classical method for determining the reaction kinetics, specifically the activation energy, is by use of the Kissinger method [27], however, many alternative methods for data analysis have been suggested and are widely utilized [28–34]. The activation energy, E_a, can be computed by plotting $\ln(\beta/T_p^2)$ as a function of $1/T_p$, where T_p is the peak temperature (**Figure 2**).

The DTA/DSC based methods are the most widely used for gasless SHS systems, with multiple studies into intermetallics, specifically the Ni/Al [35–43], Ti/Al [44–46], Co/Al [47], Al/Ru [48], Nb/Al [49], and Mg/Al [50] systems, in addition to other binary solid-solid compositions, i.e., the Si/C [51], Mo/Si [52], Zr/B [53], Fe/Se [54]. More complicated ternary systems were also investigated [54–61]. In general, a wide variety of factors can influence the measured kinetics, including variations in reactant microstructure, heating rates, among other factors. Although there are a number of studies into the same systems, it would be valuable for systematic work to be conducted where these factors are studied in depth across different systems to see what

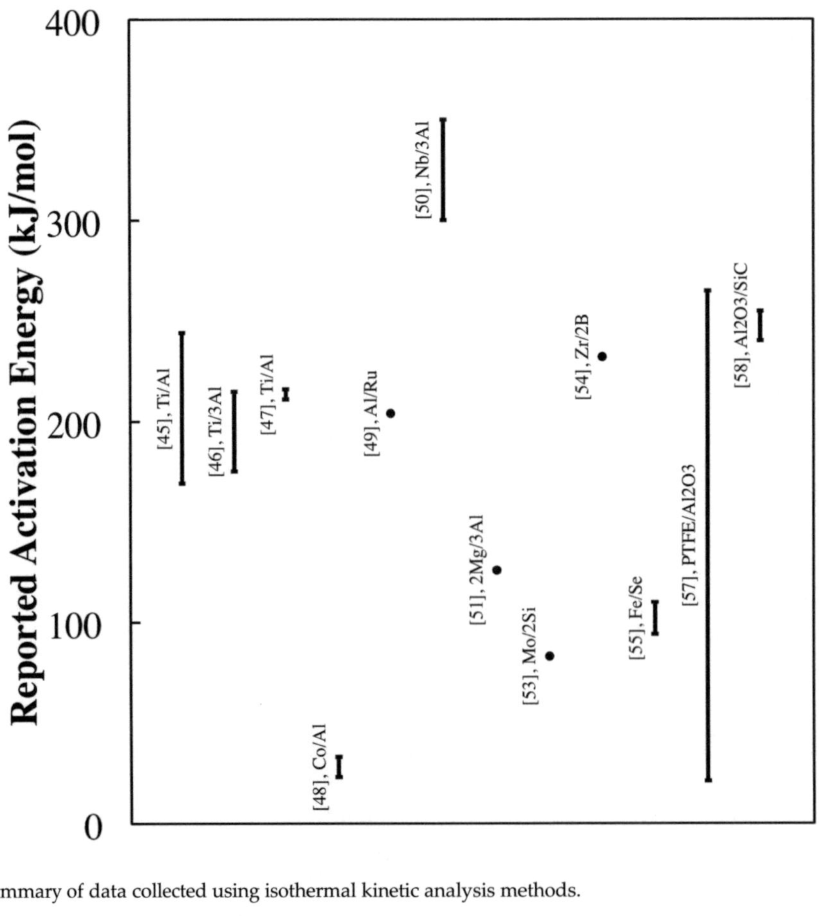

Figure 2. Summary of data collected using isothermal kinetic analysis methods.

general conclusions can be drawn. Additionally, because the experimental conditions, typically specifically heating rate and temperature ranges, are not the same as in traditional SHS, it is unclear whether the determined values can be directly compared or if there is some systemic difference that is occurring.

Figure 2 illustrates the ranges of the obtained values of activation energies for a variety of gasless exothermic reactions measured by DTA method. It can be seen that the determined values are dependent on the experimental conditions utilized, including variations in reactant microstructure, heating rates, among other factors. This issue is discussed in detail below. It would be valuable for systematic studies to be done where these factors are studied in depth across different systems. Additionally, while the activation energies reported using different DTA-based approaches does not appear to be significantly different, it is still important to understand why these differences are present and which methods of analysis are most suitable for these systems.

5. Combustion velocity/temperature analysis

There has been significant effort done to accurately correlate experimental combustion parameters, such as combustion wave velocity and temperature, with the kinetics parameters. Two major approaches have been developed to determine the activation energy by measuring the layer-by-layer combustion front combustion velocity. The first was suggested in 1977 by Merzhanov for 1D propagation [62]. The derived equation takes the form:

$$v^2 = \frac{\lambda}{(-\Delta H_r)\rho} \frac{RT_c^2}{E} \frac{k_0 \exp\left(-\frac{E}{RTc}\right)}{f(\eta_s)} \tag{8}$$

where $f(\eta_s)$ is the selected kinetic law. This technique is most commonly used by adding diluent to the sample. This affects the combustion velocity and temperature; the change in both of these values is measured then compared, leading to the kinetic relationship being understood.

The other major approach to determining the kinetics based on the velocity was developed by Boddington et al. [63]. The relationship takes the form:

$$\frac{\partial \eta}{\partial t} = \frac{\left[\frac{T-T_0}{t_d - t_r} + \frac{\partial T}{\partial t} - \left(\frac{\alpha}{v^2}\right)\left(\frac{\partial^2 T}{\partial t^2}\right)\right]}{\tau_{ad}} \tag{9}$$

where t_d and $t_{r,}$ are the decay and rise times, with τ_{ad} being the temperature rise under adiabatic conditions. More complete derivations of these two models can be found in the original articles [62–64]. Additionally, a more complete understanding of these models, including their relative merits, has been examined in a number of prior works [65–68]. In order to properly use these techniques, relatively simple equipment is required. Typically, sets of thermocouples are used to measure combustion wave propagation velocities, but there are many alternatives, such as IR or high-speed cameras, to measure the propagation velocity.

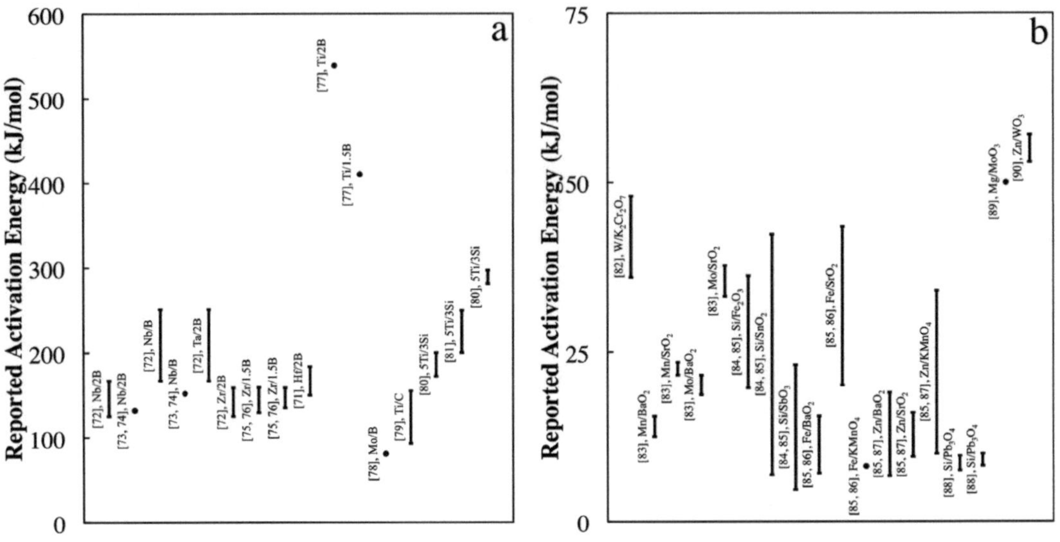

Figure 3. Summary of data collected using layer velocity analysis approached for (a) binary elemental and (b) thermite systems.

Because of the relative simplicity in using these techniques, they have found widespread use within SHS reactions. These layer velocity approaches have been used to describe the kinetics in the Ni/Al system [69, 70], in boride systems, including Nb/B [71–73], Ta/B [71], Zr/B [71, 74, 75], Hf/B [71], Ti/B [76], Mo/B [77], along with other binary systems such as Ti/C [78], Ti/Si [79, 80], thermites [81–89] and more complex ternary systems [78, 90–93].

The results obtained are presented in **Figure 3**. It can be seen that there are a wide variety of systems analyzed. It worth noting that these data were obtained through direct analysis of the combustion parameters, which were obtained at extremely high temperatures and rapid heating rates, which are difficult to accomplish by other kinetics methods. However, it is important to remember that the quasi-homogeneous approximation is utilized for this layer-by-layer combustion front combustion velocity based method, which should be applied with caution [68]. Finally, because this method can be easily used for any type of system, a comparison between thermite and non-thermite type reactions can be analyzed. Comparing **Figure 3a** and **b**, it is obvious that the activation energies for thermites is significantly lower than non-thermite type reactions. This is interesting and is likely related to the specific reaction mechanism that occurs in thermites, compared to non-thermites.

6. Electrothermography

Electrothermography is a technique that utilizes metal wires in either a gaseous or clad environment [8]. The wire is resistively heated rapidly to the desired temperature, with the electric power being adjusted to compensate for the heat release due to the chemical reaction. The obtained data allows for extraction of the rate of heat generation during the reaction under conditions similar to those in SHS wave. Additionally, because the wires are thin, the sample is quenched essentially as soon as the power is turned off. After the wire is cooled, cross-sections of the wire are collected and the width of the product films is measured, which allows for information on the kinetics of phase formation. This, when done at multiple temperatures and times, gives a more complete picture of the reaction mechanism.

The electrothermography technique has been widely used to study gas-solid reactions. This is because the wires can be exposed to any sort of gaseous environment and also because of the equipment itself; the wires can be heated with any heating rate, mirroring those found in conventional SHS reactions. Because of this, a number of experiments were conducted in nitrogen environments. In specific, the Ta/N [94], Ti/N [95–98], Nb/N [98, 99], and Zr/N [8] systems have been studies. However, the technique has also be used to study carbides, including the Ti/C [100], Zr/C [100], W/C [101], Nb/C [102], along with other systems, including Mo/Si [103–106], W/Si [107, 108], Ni/Al [109], and Ni/Ti [109].

Figure 4 shows the values of the activation energies obtained by this method in gas-solid systems. It can be seen that some reactions have limited steps relating to the gas pressure, while others are relatively independent, suggesting that there are different mechanisms for the

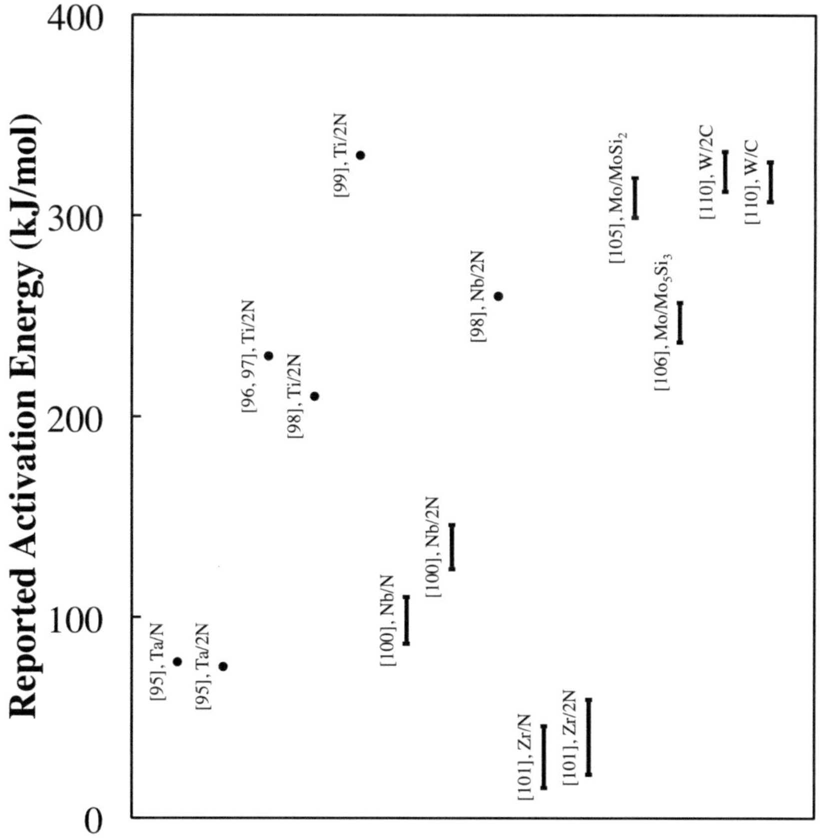

Figure 4. Summary of data obtained using the electrothermography approach.

two classes of reactions. This is also true for the gasless systems investigated. This method provides a window into determining the mechanism and, due to the nature of the experiment, allows for control over the experimental conditions to the degree that individual steps in the reaction can be isolated.

7. Modern in situ high-speed high-resolution methods

There currently exist a number of techniques to study in situ reactions on the time and length scales that occur during SHS reactions; these techniques are incredibly valuable to determine the reaction mechanisms. The most widespread technique is time-resolved X-ray diffraction (TRXRD), and is used to determine the phases that are present during the reaction. It allows for information on the phases present at every stage of the reaction, depending on the time resolution. The lower the time resolution, the more information that can be attained. Depending on the specific setup, whether synchrotron or laboratory-scale based, time resolutions ranging from 10^{-6} to 10^{-2} s are reasonable, with the absolute limit being continually improved with improved synchrotron and detector technology. There has been significant work done with SHS systems due to their solid nature, which is simple to use in TRXRD systems. It is possible to measure solid solution formations, intermediate phases, any melting processes, and the general reaction progress. Through these data it should be possible to extract kinetic data on all reaction stages based on the growth rates of the peaks for the new phase formation coupled with the decomposition of peaks from the previous phase, however, there are currently no established models illustrating this.

There have been a wide variety of experiments conducted on SHS systems by a number of different groups. For intermetallic systems, groups have studied the Ni–Al [9, 110–116], Fe–Al [111, 117–121], Nb–Al [122–124], and numerous other systems [110, 125–127]. Additionally, many groups have examined other SHS based systems, such as carbides, including Ti–C [110, 128, 129], Ta–C [129, 130], and other carbides and cermets [129, 131–134], nitrides [135, 136], oxides [137–139], silicides, including Fe–Si [140, 141], Mo–Si [119, 123, 142, 143] and Ti–Si [144, 145], among a variety of other systems [113, 129, 146–155].

In addition to TRXRD, there is a variety of other, less common, but still very useful techniques available. For example, high-speed X-ray phase-contrast imaging [10] utilizes a synchrotron source coupled with the fact that different phases absorb X-rays differently to determine which phase transformations occur during reaction, essentially high-speed X-ray phase contrast imaging. This technique was illustrated on the W-Si system at the Advanced Proton Source in Argonne National Laboratory. This method allowed for direct imaging of irreversible reactions in the W-Si reactive system at frame rates up to 36,000 frames per second with a 4-μs exposure time and spatial resolution of 10 μm. Another advanced technique is high-speed transmission electron microscopy (HSTEM) [11], which utilizes all abilities of conventional TEM, but at nanosecond time scales. This allows for direct observation of both the structural changes and crystal structure during the reaction with unprecedented resolution, as shown in **Figure 5**. Specifically, a high-time resolution dynamic transmission electron microscopy (DTEM) was developed in Lawrence Livermore National Laboratory (USA) and captures the material

http://dx.doi.org/10.5772/intechopen.70560

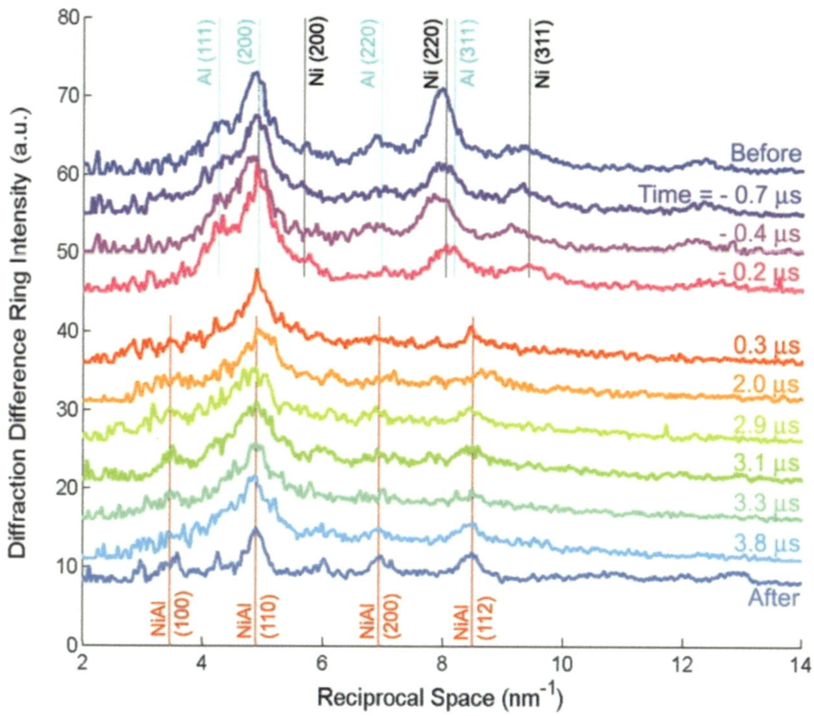

Figure 5. Dynamic single-shot diffraction with 15-ns time resolution of regions before, during, and after the exothermic mixing reaction front has passed. The times indicated at right are in relation to the reaction front, set at t = 0 s. The crystal structure clearly changes from separate fcc Al/Ni and Ni/V layers to an intermetallic B2 structure NiAl phase within 300 ns after the arrival of the hot reaction front; a.u., arbitrary units. Adapted from Kim et al. [11].

dynamics with nanosecond time resolution. The current DTEM performance shows a spatial resolution less than 10 nm for single-shot imaging, using 15 ns electron pulses. The solid-state reactions in NiAl reactive multilayer films, the martensitic transformations in nanocrystalline Ti, and the catalytic growth of Si nanowires were studied by DTEM [156].

The above unique diagnostics, which are used to determine phase and structural transformations in situ, are incredibly valuable. Used alone, they provide information on the reaction progress and mechanism; however, there are two significant paths that would make these techniques more valuable. When coupled with current methods for determination of kinetics, the understanding of the reaction mechanisms in all systems will be improved. Furthermore, there is no current way to extract kinetic parameters from some techniques (TRXRD, DTEM, etc.); it would be valuable to develop reliable approaches for these techniques.

8. Structure-kinetics relationship of SHS systems

There have been a number of studies shown that indicated that the structure of materials plays a role in the kinetics; multiple groups using a wide variety of techniques and approaches have confirmed this conclusion. For example, it was shown that by changing the internal structure of the reactive composite by using high-energy ball milling (HEBM) in the same binary system, one can significantly change the measured effective activation energy [20]. There are two approaches to quantifying this effect; the first is by rigorous quantification of the already existing structures, and the second is by use of more simple, so-called model microstructures, typically manifested in reactive nanofoils with periodically fabricated layers of reactants.

Shuck et al. utilized two techniques for quantitative determination of HEBM-produced materials [5, 157]. The first technique, X-ray Nanotomography, works by passing high brilliance X-rays through the sample and collecting the transmitted X-rays. In order to convert this 2D projection into three dimensions, the sample is rotated and the same quality of X-rays is passed through again and the projections at different angles are collected; this collection of images can be combined, leading to a 3D map of the internal sample. The second technique, focused ion beam (FIB) sectioning, uses a FIB to serially section the HEBM-produced particles. The series of images were first shear corrected, contrast normalized, and then aligned using a least-squares method, with the reconstructions shown in **Figure 6**. After structure analysis, Shuck and

Figure 6. Reconstructed internal volumes of the nanocomposites for different time (min) of WG: (a) 10, (b) 20, (c) 30, and (d) 40 (Ni is the gray phase, and Al is the void-space). Adapted from Shuck et al. [5].

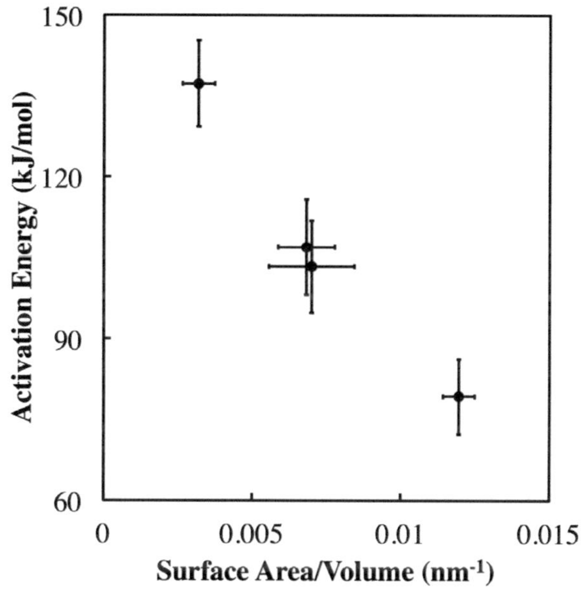

Figure 7. Dependence of effective activation energy of the reaction as a function specific contact surface area between Ni and Al phases. Adapted from Shuck and Mukasyan [20].

Mukasyan [20] further studied these effects on the kinetics in the Ni–Al system using the ETE approach. They showed, by use of the above 3D reconstruction techniques, that it is possible to control the activation energy by modification of the contact surface area between the reactants. Effectively, they lowered the effective activation energy from 79 to 137 kJ/mol, which corresponded to a change in the contact surface area/volume ratio between 0.0120 and 0.0032 nm^{-1}, respectively; this relationship is shown in **Figure 7**. Additionally, it was suggested that, for SHS systems, the apparent activation energy is affected primarily by the contributions between the diffusion and intrinsic reaction activation energies. Additionally, they offered an explanation for the relationship, relating to the difference in contribution between the diffusive activation energies (volume, grain-boundary, and surface) in conjunction with the intrinsic activation energy. This suggests that the measured and reported activation energies presented in literature are effective, or apparent, activation energies that depend highly on the structure and experimental conditions.

9. Modification of the reaction mechanism depending on the experimental conditions

In a very early study on the reaction mechanisms, Philpot et al. examined the effect of varying factors on the reaction rate [35]. In one study, they systematically varied the aluminum concentration, the heating rate, and the nickel particle size. Briefly, they showed that, depending on the applied heating rate, two different mechanisms could be initiated. The first, when using slower heating rates, was related to the melting of the aluminum metal, followed by spreading over the nickel particles. For their studies, they saw two definite peaks relating to the reaction.

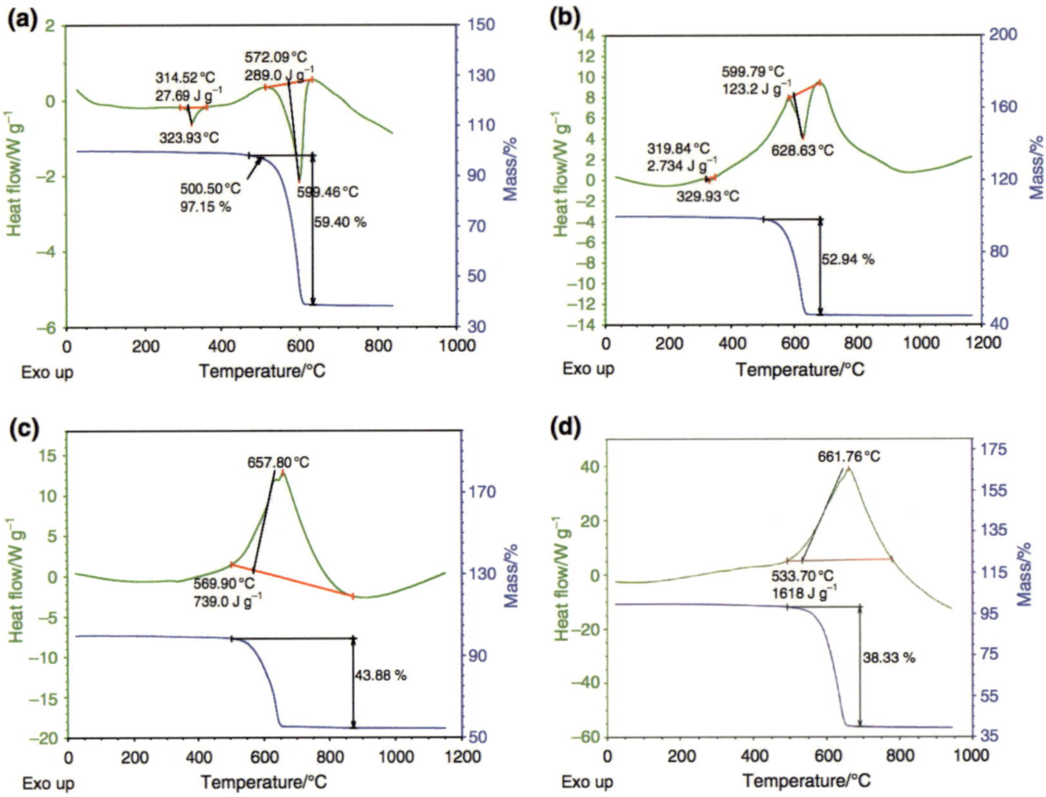

Figure 8. DTA-TG curves for the PTFE-Al_2O_3 system using heating rates of (a) 20, (b) 80, (c) 150, and (d) 160°C min^{-1} under an argon atmosphere. The additional lines are showing the changes in heat flow and mass percentage values. Adapted from Hobosyan et al. [56].

However, when they increased the heating rate, they instead only witnessed a single peak. This peak was related to the solid state transition to the final product. For transitional values between the two extremes, they found that there were relative contributions of both different mechanisms. This is an important observation which should be accounted when investigating the reaction kinetics in highly exothermic systems. Indeed, it shows that it is possible to control the reaction mechanism depending on the applied experimental conditions. This effect was confirmed by many researchers both for gasless and gas-solid systems [36, 56, 158–162] and is illustrated in **Figure 8**. Furthermore, depending on the reaction mechanism, it is possible to control both the final product phases and their microstructures, thus producing materials with tailored properties.

10. The Ni/Al system as a model for SHS kinetics

In the SHS community, the Ni/Al system has been widely used a model system. It was chosen because of its low ignition temperature, high oxidation resistance, and ease of processing. Because of this, all of the above mentioned techniques have been utilized to study kinetics in this system. Thus it is possible to compare the data collected across a large number of experimental conditions to give more complete understanding of this gasless reaction. Although this system has been extensively studied for over 40 years now, a consensus has not emerged on the exact activation energy, as can be seen in **Figure 9**. The data again illustrate the effect that differing experimental conditions play, whether in the material structure, heating rates, or other experimental factors.

In 1987, Philpot et al. did preliminary studies on the Ni/Al system kinetics using the Kissinger approach [35]. They showed that there is an effect of heating rate, the nickel particle size, as well as an effect of a varying Al content, on the reaction mechanism in this system [35], this work is highlighted in more detail in an above section. Hunt et al. examined the effect of particle size on the apparent activation energy using a CO_2 laser to control the heating coupled with the DTA-based Kissinger approach. Using 800 nm size Ni particles coupled with 20 μm, 4 μm, 80 nm, and 40 nm Al particles, it was found that that these reactions had activation energies of: 103.5 ± 10.5, 97.3 ± 5, 21.2 ± 2.5, and 17.4 ± 2.85 kJ/mol, respectively [36]. They then increased the Ni size to 15 μm and measured with the four Al sizes, resulting in 162.5 ± 1.4, 131.2 ± 2.6, 103.6 ± 5.2, and 80.1 ± 6.3 kJ/mol, respectively [36]. This confirms that the initial reactant structure plays a significant impact on the kinetics. The size of either reactant significantly alters the effective kinetics. Kim et al. studied this reaction using nanolaminated composite micro-foils with different thickness ratios between Ni and Al at heating rates between 5 and 100 K/min. For 4:1 foils, they found that the formation of $NiAl_3$ occurred, followed by Ni_2Al_3, with activation energies of 142 and 106 kJ/mol, respectively [37].

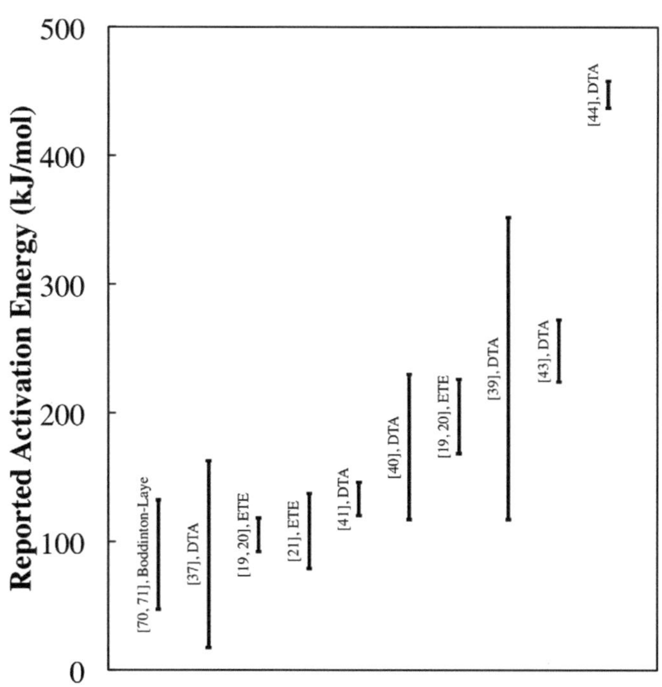

Figure 9. Summary of kinetic data collected for the Ni/Al system.

In order to more fully understand the relationship between structure and the kinetics, White et al. investigated the effect of mechanical activation (MA) on the Ni/Al system kinetics using the Kissinger approach. Two types of Ni/Al composites were used; Ni clad Al particles, as well as Ni/Al composite particles produced by high-energy ball milling (HEBM). The Ni clad Al particles were found to have an apparent activation energy of 352 ± 8 kJ/mol, while after MA the particles had much lower activation energy of 117 ± 4 kJ/mol [38]. Reeves et al. studied the thermal and impact reaction kinetics in the Ni/Al system for both MA and nano-sized reactants using the Kissinger approach. For the nano-mixture, the reactants were both ~80 nm in size and, for the MA particles, they underwent 15 minutes of HEBM. The nano-mixture exhibited a 230 ± 21 kJ/mol activation energy, while the MA mixture was calculated to be 117 ± 8 kJ/mol [39]. Manukyan et al. studied the Ni/Al system after MA and the effect of a coarse vs. nanolaminated nanostructure on the kinetics using the Kissinger approach. Using heating rates between 10 and 50 K/min, they found that the reaction proceeds in three steps, $NiAl_3$, Ni_2Al_3, and then finally NiAl. For the coarse microstructure, these peaks corresponded to 99 ± 4, 138 ± 13, and 120 ± 37 kJ/mol, while the nanolaminated microstructure corresponded to 93 ± 2.5, 145 ± 13, and 146 ± 14 kJ/mol, respectively [40]. This illustrated that the activation energies depend on the microstructure, even after MA.

Kuk et al. studied compression bonded Ni/Al nanofoils with and without a BN lubricant using the Kissinger approach. With the BN lubricant, it was found that the reaction proceeded in two steps, with the activation energies being 224 and 272 kJ/mol, respectively, resulting in the formation of Al_3Ni_2 [42]. Without the lubricant, the reaction proceeded in a single step with activation energy of 470 kJ/mol, this difference was attributed to the oxide layer between the reactants [42]. Maiti and Ghoroi studied the Ni/Al system using the Friedman, Ozawa, and Kissinger approach, yielding activation energies of 437.0, 448.4, and 457.6 kJ/mol, respectively [43].

Using the ETE approach, Shteinberg et al. and Mukasyan et al. confirmed that MA affects the activation energy in the Ni/Al system, as shown in **Figure 10** [12, 19]. Initially, they studied the kinetics of the Ni clad Al system (which was also studied in [38]), which showed two distinct steps. The first step was related to the melting of Al and subsequent cracking of the Ni layer, which had an activation energy of 197 ± 29 kJ/mol. The final step was the diffusion of Ni into Al, which was measured as 523 ± 84 kJ/mol. In the MA system, they found that only a single step occurred and was measured to be 105 ± 13 kJ/mol. Shuck and Mukasyan further studied the effects of MA on the kinetics in the Ni/Al system using the ETE approach [20]. Using 3D reconstruction techniques, they showed that the surface area contact between the reactions is directly related to the effective activation energy, which ranged from 79 to 137 kJ/mol, which corresponded to a change in the contact surface area/volume ratio between 0.0120 and 0.0032 nm^{-1}, respectively. This work is further highlighted in an above section. Finally, using the ETE approach, Filimonov et al. studied the effects of MA on the nonstoichiometric, 3Ni/Al system [21]. They utilized a criterion based on the minimum curvature of the heating rate logarithm, which resulted in an anomalously low measured activation energy of 9.5 ± 2 kJ/mol.

Marin-Ayral et al. studied the Ni/Al system under different gas pressures using the Boddington-Laye method. They showed that for pressures of 100, 320, and 500 MPa, the measured activation energies were 47, 59, and 132 kJ/mol, respectively [68, 69]. Vadchenko et al. studied the Ni/Al

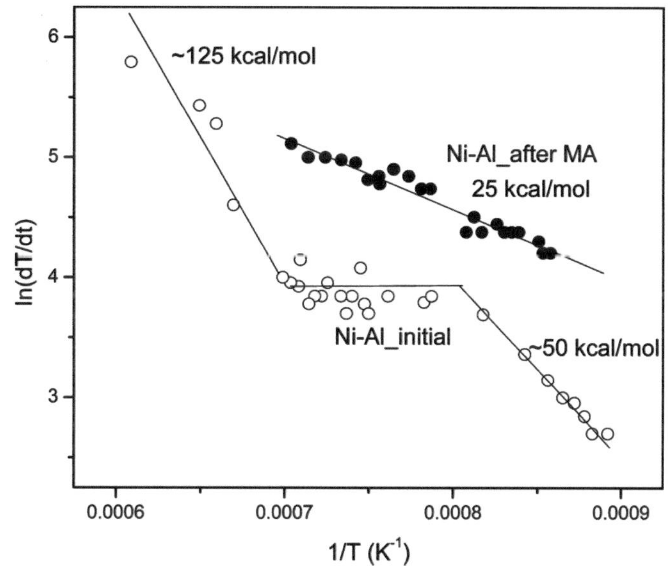

Figure 10. Arrhenius plots for reactions in Al clad by Ni systems before and after high energy ball milling. Adapted from Shteinberg et al. [12].

system using electrothermography. Their results showed that the reactions occur first through grain boundary diffusion, followed by diffusion of the solid metal into the liquid phase [109]. Finally, Mukasyan et al. examined the Ni/Al system using a combination of TRXRD and ETE, showing that the reaction mechanism itself changes based on the structure, this work is highlighted in an above section [116].

Thus we may conclude that although a wide variety of studies were conducted on the Ni/Al system with many different experimental and structural conditions, the reported values of activation energies vary drastically. Additionally, there would be great benefit to combining utilizing multiple methods simultaneously to bridge understanding between the different experimental techniques.

11. Future directions in SHS kinetics

More complete understanding on the kinetics of SHS reactions is vital for both fundamental science and also for practical or industrial reasons. To better understand the kinetics, combinations of techniques must be utilized, specifically coupling techniques that give information on the kinetics while simultaneously examining the phase transformations that are occurring. To further understand the reaction mechanisms, additional studies must be conducted on the relationship between the structure and the resulting kinetics. Additionally, work must be done to compare the different methods of studying kinetics and their interrelationships. In the limited cases where there is data available for the same system across different techniques, there is a wide range of published kinetic data. It is imperative to continue to study SHS kinetics in a more systematic, fundamental fashion.

Acknowledgements

This work was supported by the Department of Energy, National Nuclear Security Administration, under the award number DE-NA0002377 as part of the Predictive Science Academic Alliance Program II. We also acknowledge the Ministry of Education and Science of the Russian Federation specifically Increase Competitiveness Program of NUST 'MISiS' (No. K2-2016-065), implemented by a governmental decree dated 16th of March 2013, N 211. Finally, this work was also supported by the U.S. Department of State through the Fulbright program.

Author details

Christopher E. Shuck[1] and Alexander S. Mukasyan[1,2]*

*Address all correspondence to: amoukasi@nd.edu

1 Department of Chemical and Biomolecular Engineering, University of Notre Dame, Notre Dame, Indiana, United States

2 University of Notre Dame and National University of Science and Technology MISiS, Moscow, Russia

References

[1] Merzhanov AG, Borovinskaya IP. Self-spreading high-temperature synthesis of refractory compounds. Doklady Chemistry. 1972;**204**(2):429-431

[2] Rogachev AS, Mukasyan AS. Combustion for Materials Synthesis. CRC Press, Taylor and Francis; 2015

[3] Gorban AN, Yablonsk GS. Three waves of chemical dynamics. Mathematical Modelling of Natural Phenomena. 2015;**10**(5):1-5

[4] Shuck CE, Manukyan KV, Rouvimov S, Rogachev AS, Mukasyan AS. Solid-flame: Experimental validation. Combustion and Flame. 2016;**162**:487-493

[5] Shuck CE, Pauls JM, Mukasyan AS. Ni/Al energetic nanocomposites and the solid flame phenomenon. Journal of Physical Chemistry C. 2016;**120**(47):27066-27078

[6] Šimon P. Isoconversional methods. Journal of Thermal Analysis and Calorimetry. 2004; **76**(1):123-132

[7] Merzhanov AG, Borovinskaya IP, Shteinberg AS, Kochetov OA, Ulybin VB, Shipilov VV. Method for production of the refractory compounds, 2130084/22-02. Russian patent: Bulletin Izobreteniy. 1977;**46**:64

[8] Merzhanov AG, Grigorev YM, Kharatyan SL, Mashkinov LB, Vartanyan ZS. Study on heat-evolution kinetics at high-temperature nitration of zirconium wires. Fizika Goreniya i Vzryva. 1975;**11**(4):563-568

[9] Boldyrev VV, Aleksandrov VV, Korchagin MA, Tolochko BP, Guzenko SN, Sokolov AS, Sheromov MA, Lyakhov NZ. Study of phase formation dynamics at nickel monoaluminide synthesis in combustion regime. Doklady Akademii Nauk SSSR. 1981;**259**(5):1127-1129

[10] Reeves RV, White JDE, Dufresne EM, Fezzaa K, Son SF, Varma A, Mukasyan AS. Microstructural transformations and kinetics of high-temperature heterogeneous gasless reactions by high-speed X-ray phase-contrast imaging. Physical Review B. 2009;**80**:224103

[11] Kim JS, LaGrange T, Reed BW, Taheri ML, Armstrong MR, King WE, Browning ND, Campbell GH. Imaging of transient structures using nanosecond in situ TEM. Science. 2008;**321**(5895):1472-1475

[12] Shteinberg AS, Lin YC, Son SF, Mukasyan AS. Kinetics of high temperature reaction in Ni-Al system: Influence of mechanical activation. The Journal of Physical Chemistry. A. 2010;**114**(20):6111-6116

[13] Knyazik VA, Merzhanov AG, Solomonov VB, Shteinberg AS. Macrokinetics of high-temperature titanium interaction with carbon under electrothermal explosion conditions. Combustion, Explosion and Shock Waves. 1985;**21**(3):333-337

[14] Knyazik V, Shteinberg A, Gorovenko V. Thermal analysis of high-speed high-temperature reactions of refractory carbide synthesis. Journal of Thermal Analysis and Calorimetry. 1993;**40**(1):363-371

[15] Kostogorova JY, Viljoen HJ, Shteinberg AS. Study of the high-temperature solid-phase titanium-carbon reaction. Industrial & Engineering Chemistry Research. 2003;**42**(26):6714-6719

[16] Knyazik V. High-temperature interaction in the Ta-C system under electrothermal explosion conditions. Journal of Materials Synthesis and Processing. 1993;**1**(2):85-92

[17] Gorovenko VI, Knyazik VA, Shteinberg AS. High-temperature interaction between silicon and carbon. Ceramics International. 1993;**19**(2):129-132

[18] Popov KV, Knyazik VA, Shteinberg AS. Study of high-temperature reaction of Ti with B by the method of electrothermal explosion. Combustion, Explosion and Shock Waves. 1993;**29**(1):77-81

[19] Mukasyan AS, Khina BB, Reeves RV, Son SF. Mechanical activation and gasless explosion: Nanostructural aspects. Chemical Engineering Journal. 2011;**174**(2):677-686

[20] Shuck CE, Mukasyan AS. Reactive Ni/Al nanocomposites: Structural characteristics and activation energy. The Journal of Physical Chemistry. A. 2017;**121**(6):1175-1181

[21] Filimonov VY, Korchagin MA, Smirnov EV, Lyakhov NZ. Macrokinetics of solid-phase synthesis of an activated 3Ni+ Al mixture in the thermal explosion mode. Combustion, Explosion and Shock Waves. 2010;**46**(4):449-456

[22] Lin YC, Shteinberg AS, McGinn PJ, Mukasyan AS. Kinetics study in Ti-Fe_2O_3 system by electro-thermal explosion method. International Journal of Thermal Sciences. 2014;**84**: 369-378

[23] Shilyaev MI, Borzykh VÉ, Dorokhov AR, Ovcharenko VE. Determination of thermokinetic parameters from the inverse problem of an electrothermal explosion. Combustion, Explosion and Shock Waves. 1992;**28**(3):258-262

[24] Bostandzhiyan S, Gordopolova I, Shcherbakov V. Modeling of an electrothermal explosion in gasless systems placed into an electroconducting medium. Combustion, Explosion and Shock Waves. 2013;**49**(6):668-675

[25] Bostandzhiyan SA, Gordopolova IS, Shcherbakov AV, Shcherbakov VA. Electrothermal explosion in cylindrical Ti-C charges covered with a TiC shell: A mathematical model. International Journal of Self-Propagating High-Temperature Synthesis. 2012;**21**(3):183-188

[26] Bostandzhiyan SA, Shcherbakov AV, Shcherbakov VA. Mathematical modeling of electrothermal explosion in gasless systems placed in a hollow dielectric cylinder. International Journal of Self-Propagating High-Temperature Synthesis. 2016;**25**(2):75-79

[27] Kissinger HE. Reaction kinetics in differential thermal analysis. Analytical Chemistry. 1957;**29**(11):1702-1706

[28] Borchardt HJ, Daniels F. The application of differential thermal analysis to the study of reaction kinetics. Journal of the American Chemical Society. 1957;**79**(1):41-46

[29] Friedman HL. Kinetics of thermal degradation of char-forming plastics from thermogravimetry. Application to a phenolic plastic. Journal of Polymer Science: Polymer Symposia. 1964;**6**(1):183-195

[30] Ozawa T. A quick, a new method of analyzing thermogravimetric data. Bulletin of the Chemical Society of Japan. 1965;**38**(11):1881-1886

[31] Flynn JH, Wall LA. Direct method for the determination of activation energy from thermogravimetric data. Journal of Polymer Science Part C: Polymer Letters. 1966;**4**(5): 323-328

[32] Ozawa T. Kinetics of non-isothermal crystallization. Polymer. 1971;**12**(3):150-158

[33] Starink MJ. A new method for the derivation of activation energies from experiments performed at constant heating rate. Thermochimica Acta. 1996;**288**(1–2):97-104

[34] Vyazovkin S, Wight CA. Model-free and model-fitting approaches to kinetic analysis of isothermal and nonisothermal data. Thermochimica Acta. 1999;**340**:53-68

[35] Philpot KA, Munir ZA, Holt JB. An investigation of the synthesis of nickel aluminides through gasless combustion. Journal of Materials Science. 1987;**22**(1):159-169

[36] Hunt EM, Pantoya ML. Ignition dynamics and activation energies of metallic thermites: from nano-to micron-scale particulate composites. Journal of Applied Physics. 2005;**98**(3): 034909

[37] Kim HY, Chung DS, Hong SH. Reaction synthesis and microstructures of NiAl/Ni micro-laminated composites. Materials Science and Engineering: A. 2005;**396**(1):376-384

[38] White JD, Reeves RV, Son SF, Mukasyan AS. Thermal explosion in Al-Ni system: Influence of mechanical activation. Journal of Physical Chemistry C. 2009;**113**(48):13541-13547

[39] Reeves RV, Mukasyan AS, Son SF. Journal of Physical Chemistry C. 2010;**114**(35):14772-14780

[40] Manukyan KV, Mason BA, Groven LJ, Lin YC, Cherukara M, Son SF, Strachan A, Mukasyan AS. Tailored reactivity of Ni+ Al nanocomposites: Microstructural correlations. Journal of Physical Chemistry C. 2012;**116**(39):21027-21038

[41] Baghdasaryan AM, Hobosyan MA, Khachatryan HL, Niazyan OM, Kharatyan SL, Sloyan LH, Grigoryan YG. The role of chemical activation on the combustion and phase formation laws in the Ni-Al-promoter system. Chemical Engineering Journal. 2002;**188**:210-215

[42] Kuk SW, Yu J, Ryu HJ. Effects of interfacial Al oxide layers: Control of reaction behavior in micrometer-scale Al/Ni multilayers. Materials & Design. 2015;**84**:372-377

[43] Maiti SC, Ghoroi C. Thermo-kinetic analysis of Ni -al intermetallic phase formation in powder system. Journal of Thermal Analysis and Calorimetry. 2016;**124**(2):1039-1051

[44] Illeková E, Gachon JC, Rogachev A, Grigoryan H, Schuster JC, Nosyrev A, Tsygankov P. Kinetics of intermetallic phase formation in the Ti/Al multilayers. Thermochimica Acta. 2008;**469**(1):77-85

[45] Sina H, Iyengar S. Reactive synthesis and characterization of titanium aluminides produced from elemental powder mixtures. Journal of Thermal Analysis and Calorimetry. 2015;**122**(2):689-698

[46] Adeli M, Seyedein SH, Aboutalebi MR, Kobashi M, Kanetake N. Implementation of DSC analysis in reaction kinetics during heating of Ti-50 at.% Al powder mixture. Journal of Thermal Analysis and Calorimetry. 2017;**128**(2):867-874

[47] Reeves RV, Adams DP. Reaction instabilities in Co/Al nanolaminates due to chemical kinetics variation over micron-scales. Journal of Applied Physics. 2014;**115**(4):044911

[48] Pauly C, Woll K, Bax B, Mücklich F. The role of transitional phase formation during ignition of reactive multilayers. Applied Physics Letters. 2015;**107**(11):113104

[49] Mostaan H, Karimzadeh F, Abbasi MH. Synthesis and formation mechanism of nano-structured NbAl 3 intermetallic during mechanical alloying and a kinetic study on its formation. Thermochimica Acta. 2012;**529**:36-44

[50] Zhou Y, Zhu Y, Zhu Y, Li L. Phase transformation, kinetics and thermodynamics during the combustion synthesis of Mg_2Al_3 alloy. Journal of Alloys and Compounds. 2015;**628**:257-262

[51] Hu Q, Zhang M, Luo P, Song M, Li J. Thermal explosion synthesis of ZrC particles and their mechanism of formation from Al-Zr-C elemental powders. International Journal of Refractory Metals and Hard Materials. 2012;**35**:251-256

[52] Takacs L, Soika V, Baláž P. The effect of mechanical activation on highly exothermic powder mixtures. Solid State Ionics. 2001;**141**:641-647

[53] Lee D, Sim GD, Xiao K, Vlassak JJ. Low-temperature synthesis of ultra-high-temperature coatings of ZrB_2 using reactive multilayers. Journal of Physical Chemistry C. 2014;**118**(36):21192-21198

[54] Liu G, Li J, Chen K. Reaction mechanism in fast combustion synthesis of superconducting FeSe and FeSe0.7Te0.3. Acta Materialia. 2017;**122**:187-198

[55] Kim BG, Lee YW, Lee JW, Choi Y. Formation mechanism of C/Sic/C multi-layers during self-propagating high-temperature synthesis. Surface and Coatings Technology. 2002; **151**:26-30

[56] Hobosyan MA, Kirakosyan KG, Kharatyan SL, Martirosyan KS. PTFE -Al_2O_3 reactive interaction at high heating rates. Journal of Thermal Analysis and Calorimetry. 2015;**119**(1): 245-251

[57] Pathak LC, Bandyopadhyay D, Srikanth S, Das SK, Ramachandrarao P. Effect of heating rates on the synthesis of Al_2O_3-SiC composites by the self-propagating high-temperature synthesis (SHS) technique. Journal of the American Ceramic Society. 2001;**84**(5):915-920

[58] Choi Y, Rhee SW. Reaction of TiO_2-Al-C in the combustion synthesis of TiC-Al_2O_3 composite. Journal of the American Ceramic Society. 1995;**78**(4):986-992

[59] Zhu H, Jiang Y, Yao Y, Song J, Li J, Xie Z. Reaction pathways, activation energies and mechanical properties of hybrid composites synthesized in-situ from Al-TiO_2-C powder mixtures. Materials Chemistry and Physics. 2012;**137**(2):532-542

[60] Rafiei M, Enayati MH, Karimzadeh F. Kinetic analysis of thermite reaction in Al-Ti-Fe_2O_3 system to produce (Fe, Ti) 3Al-Al_2O_3 nanocomposite. Powder Technology. 2014;**253**:553-560

[61] Jordan JL, Thadhani NN. Effect of shock-activation on post-shock reaction synthesis of ternary ceramics. AIP Conference Proceedings. 2002;**620**(1):1097-1100

[62] Merzhanov AG. New elementary combustion models of 2nd kind. Doklady Akademii Nauk SSSR. 1977;**233**(6):1130-1133

[63] Boddington T, Laye PG, Tipping J, Whalley D. Kinetic analysis of temperature profiles of pyrotechnic systems. Combustion and Flame. 1986;**63**(3):359-368

[64] Boddington T, Cottrell A, Laye PG. Combustion transfer in gasless pyrotechnics. Combustion and Flame. 1990;**79**(3–4):234-241

[65] Laye PG, Constantinou CP, Volk F. Experimental studies of the propagation of combustion in solids [and discussion]. Philosophical Transactions of the Royal Society A. 1992;**339**(1654):387-394

[66] Varma A, Lebrat JP. Combustion synthesis of advanced materials. Chemical Engineering Science. 1992;**47**(9–11):2179-2194

[67] Baras F. Determination of transport and kinetic properties in self-propagating high-temperature synthesis. Journal of Alloys and Compounds. 2008;**455**(1):113-120

[68] Mukasyan AS, Rogachev AS. Discrete reaction waves: Gasless combustion of solid powder mixtures. Progress in Energy and Combustion Science. 2008;**34**(3):377-416

[69] Dumez MC, Marin-Ayral RM, Tédenac JC. The role of experimental parameters in combustion synthesis of NiAl under high gas pressure. Journal of Alloys and Compounds. 1998;**268**(1):141-151

[70] Marin-Ayral RM, Dumez MC, Tédenac JC. Influence of high gas pressure on combustion synthesis of the solid-solid reaction of NiAl compound. Materials Research Bulletin. 2000;**35**(2):233-243

[71] Zenin AA, Merzhanov AG, Nersisyan GA. Thermal wave structure in SHS processes. Combustion, Explosion and Shock Waves. 1981;**17**(1):63-71

[72] Yeh CL, Chen WH. Preparation of niobium borides NbB and NbB_2 by self-propagating combustion synthesis. Journal of Alloys and Compounds. 2006;**420**(1):111-116

[73] Yeh CL, Chen WH. A comparative study on combustion synthesis of Nb-B compounds. Journal of Alloys and Compounds. 2006;**422**(1):78-85

[74] Munir ZA. Reaction synthesis processes: Mechanisms and characteristics. Metallurgical Transactions A. 1992;**23**(1):7-13

[75] Dunmead SD, Munir ZA, Holt JB. Temperature profile analysis of combustion in the Zr-B system using the Boddington-Laye method. International Journal of Self-Propagating High-Temperature Synthesis. 1992;**1**:22-32

[76] Holt JB, Kingman DD, Bianchini GM. Kinetics of the combustion synthesis of TiB_2. Materials Science and Engineering. 1985;**71**:321-327

[77] Yeh CL, Hsu WS. Preparation of MoB and MoB-$MoSi_2$ composites by combustion synthesis in SHS mode. Journal of Alloys and Compounds. 2007;**440**(1):193-198

[78] Dunmead SD, Readey DW, Semler CE, Hol JB. Journal of the American Ceramic Society. 1989;**72**(12):2318-2324

[79] Wang LL, Munir ZA. Metallurgical and Materials Transactions B. 1995;**26**(3):595-601

[80] Kachelmyer CR, Varma A, Rogachev AS, Sytschev AE. Kinetics of combustion synthesis in the Ti-C and Ti-C-Ni systems. Industrial & Engineering Chemistry Research. 1998;**37**(6): 2246-2249

[81] Boddington T, Laye PO. Temperature dependence of the burning velocity of gasless pyrotechnics. Thermochimica Acta. 1987;**120**:203-206

[82] Drennan RL, Brown ME. Binary and ternary pyrotechnic systems of Mn and/or Mo and BaO_2 and/or SrO_2: Part 3. Kinetic aspects. Thermochimica Acta. 1992;**208**:247-259

[83] Anil Rugunanan R, Brown ME. Combustion of binary and ternary silicon/oxidant pyrotechnic systems, part IV: Kinetic aspects. Combustion Science and Technology. 1993;**95**(1–6):117-138

[84] Brown M. Some thermal studies on pyrotechnic compositions. Journal of Thermal Analysis and Calorimetry. 2001;**65**(2):323-334

[85] Tribelhorn MJ, Blenkinsop MG, Brown ME. Combustion of some iron-fuelled binary pyrotechnic systems. Thermochimica Acta. 1995;**256**(2):291-307

[86] Tribelhorn MJ, Venables DS, Brown ME. Combustion of some zinc-fuelled binary pyrotechnic systems. Thermochimica Acta. 1995;**256**(2):309-324

[87] Jakubko J. Combustion of the silicon-red lead system. Temperature of burning, kinetic analysis and mathematical model. Combustion Science and Technology. 1999;**146**(1–6): 37-55

[88] Manukyan KV, Kirakosyan KG, Grigoryan YG, Niazyan OM, Yeghishyan AV, Kirakosyan AG, Kharatyan SL. Mechanism of molten-salt-controlled thermite reactions. Industrial & Engineering Chemistry Research. 2011;**50**(19):10982-10988

[89] Won HI, Nersisyan HH, Won CW. A macrokinetic study of the WO_3/Zn reaction diluted with NaCl. Chemical Engineering Journal. 2009;**153**(1):193-198

[90] LaSalvia JC, Meyers MA. Combustion synthesis in the Ti-C-Ni-Mo system: Part II, analysis. Metallurgical and Materials Transactions A. 1995;**26**(11):3011-3019

[91] Tomasi R, Munir ZA. Effect of particle size on the reaction wave propagation in the combustion synthesis of Al_2O_3-ZrO_2-Nb composites. Journal of the American Ceramic Society. 1999;**82**(8):1985-1992

[92] Marinšek M, Kemperl J, Likozar B, Macek J. Temperature profile analysis of the citrate-nitrate combustion system. Industrial & Engineering Chemistry Research. 2008;**47**(13): 4379-4386

[93] Marinšek M. Analysis of the temperature profiles during the combustion synthesis of doped lanthanum gallate. Materiali in tehnologije. 2008;**42**(2):85-91

[94] Vadchenko SG, Grigoriev YM, Merzhanov AG. The kinetics of high-temperature tantalum nitriding. Bulletin Akademii Nauk USSR Metals. 1980;**5**:223-229

[95] Pelekh AE, Mukasyan AS, Varma A. Kinetics of rapid high-temperature reactions: Titanium-nitrogen system. Industrial & Engineering Chemistry Research. 1999;**38**(3): 793-798

[96] Pelekh A, Mukasyan A, Varma A. Electrothermography apparatus for kinetics of rapid high-temperature reactions. The Review of Scientific Instruments. 2000;**71**(1):220-223

[97] Thiers L, Leitenberger BJ, Mukasyan AS, Varma A. Influence of preheating rate on kinetics of high-temperature gas-solid reactions. AIChE Journal. 2000;**46**(12):2518-2524

[98] Thiers L, Mukasyan AS, Pelekh A, Varma A. Kinetics of high-temperature reaction in titanium-nitrogen system: Nonisothermal conditions. Chemical Engineering Journal. 2001;**82**(1):303-310

[99] Vadchenko SG, Grigor'ev YM. Investigation of kinetics and mechanism of high temperature nitridation of niobium. Izvestiya Akademii Nauk USSR Metallurgy. 1979;187-195

[100] Vadchenko SG, Grigorev YM, Merzhanov AG. Non-isothermal kinetics and mechanism of tungsten siliconizing in gasless combustion wave. Combustion, Explosion and Shock Waves. 1976;**12**(5):606-612

[101] Kharatyan SL, Chatilyan HA, Arakelyan LH. Regularities of heat release in tungsten siliconizing in a gasless combustion wave. Materials Research Bulletin. 2008;**43**(4):897-906

[102] Adamyan TA, Kharatyan SL. Reaction diffusion in Mo-Si system above the melting point of silicon. Journal of Alloys and Compounds. 2010;**496**(1):418-422

[103] Galstyan GS, Chatilyan HA, Kirakosyan AG, Kharatyan SL, Mukasyan AS, Varma A. Diffusion annealing of Mo/$MoSi_2$ couple and silicon diffusivity in Mo_5Si_3 layer. Defect and Diffusion Forum. 2005;**237**:873-878

[104] Chatilyan HA, Kharatyan SL, Harutyunyan AB. Growth kinetics of Mo_3Si layer in the Mo_5Si_3/Mo diffusion couple. Materials Science and Engineering: A. 2007;**459**(1):227-232

[105] Kharatyan SL, Chatilyan HA, Galstyan GS. Non-isothermal phenomena in Mo/Si diffusion couple: Reaction kinetics and structure formation. Thin Solid Films. 2008;**516**(15): 4876-4881

[106] Kharatyan SL, Chatilyan HA, Aghayan MA, Rodriguez MA. Investigation of the mechanism of the ignition and combustion of the systems Ti+ C, Zr+ C by an electrothermographic method. International Journal of Self-Propagating High-Temperature Synthesis. 2013;**22**(1):18-26

[107] Kharatyan SL, Chatilyan HA. On the singularity of high temperature carbidization of niobium. International Journal of Self-Propagating High-Temperature Synthesis. 1999; **8**:31-42

[108] Kharatyan SL, Chatilyan AA. Kinetics of tungsten carbidization under non-isothermal conditions. Combustion, Explosion and Shock Waves. 2000;**36**(3):342-348

[109] Vadchenko SG, Bulaev AM, Gal'chenko YA, Merzhanov AG. Interaction mechanism in laminar bimetal nickel-titanium and nickel-aluminum systems. Combustion, Explosion and Shock Waves. 1987;**23**(6):706-715

[110] Wong J, Larson EM, Holt JB, Waide PA, Rupp B, Frahm R. Time-resolved X-ray diffraction study of solid combustion reactions. Science. 1990;**249**:1406-1410

[111] Bernard F, Gaffet E, Gramond M, Gailhanou M, Gachon JC. Simultaneous Ir and time-resolved X-ray diffraction measurements for studying self-sustained reactions. Journal of Synchrotron Radiation. 2000;**7**(1):27-33

[112] Curfs C, Turrillas X, Vaughan GBM, Terry AE, Kvick Å, Rodríguez MA. Al-Ni intermetallics obtained by SHS: A time-resolved X-ray diffraction study. Intermetallics. 2007;**15**(9): 1163-1171

[113] Tingaud D, Stuppfler L, Paris S, Vrel D, Bernard F, Penot C, Nardou F. Time-resolved X-ray diffraction study of SHS-produced NiAl and NiAl-ZrO_2 composites. International Journal of Self-Propagating High-Temperature Synthesis. 2007;**16**(1):12-17

[114] Trenkle JC, Koerner LJ, Tate MW, Gruner SM, Weihs TP, Hufnagel TC. Phase transformations during rapid heating of Al/Ni multilayer foils. Applied Physics Letters. 2008;**93**(8):081903

[115] Kovalev DY, Kochetov NA, Ponomarev VI, Mukasyan AS. Effect of mechanical activation on thermal explosion in Ni-Al mixtures. International Journal of Self-Propagating High-Temperature Synthesis. 2010;**19**(2):120-125

[116] Mukasyan AS, White JDE, Kovalev DY, Kochetov NA, Ponomarev VI, Son SF. Dynamics of phase transformation during thermal explosion in the Al-Ni system: Influence of mechanical activation. Physica B. 2010;**405**(2):778-784

[117] Gaffet E, Charlot F, Klein D, Bernard F, Niepce JC. Mechanically activated SHS reaction in the Fe-Al system: In situ time resolved diffraction using synchrotron radiation. Materials Science Forum. 1998;**269**:379-384

[118] Bernard F, Charlot F, Gaffet E, Niepce JC. Optimization of MASHS parameters to obtain a nanometric FeAl intermetallic. International Journal of Self-Propagating High-Temperature Synthesis. 1998;**7**:233-248

[119] Charlot F, Gras C, Gramond M, Gaffet E, Bernard F, Niepce JC. Développements Récents De L'étude En Temps Réel Par Diffraction Des Rayons X Couplée À Une Thermographie Infrarouge: Application Au Suivi De La Réaction MASHS Dans Les Systèmes FeAl, Et $MoSi_2$. Le Journal de Physique IV. 1998;**8**(PR5):497-504

[120] Charlot F, Bernard F, Gaffet E, Klein D, Niepce JC. In situ time-resolved diffraction coupled with a thermal Ir camera to study mechanically activated SHS reaction: Case of Fe-Al binary system. Acta Materialia. 1998;**47**(2):619-629

[121] Vrel D, Girodon-Boulandet N, Paris S, Mazue JF, Couqueberg E, Gailhanou M, Thiaudiere D, Gaffet E, Bernard F. A new experimental setup for the time resolved

X-ray diffraction study of self-propagating high-temperature synthesis. The Review of Scientific Instruments. 2002;**73**(2):422-428

[122] Gauthier V, Bernard F, Gaffet E, Josse C, Larpin JP. In-situ time resolved X-ray diffraction study of the formation of the nanocrystalline $NbAl_3$ phase by mechanically activated self-propagating high-temperature synthesis reaction. Materials Science and Engineering: A. 1999;**272**(2):334-341

[123] Bernard F, Charlot F, Gras C, Gauthier V, Gaffet E. In-situ time-resolved X-ray diffraction experiments applied to self-sustained reactions from mechanically activated mixtures. Le Journal de Physique IV. 2002;**10**:89-99

[124] Gauthier V, Bernard F, Gaffet E, Vrel D, Gailhanou M, Larpin JP. Investigations of the formation mechanism of nanostructured $NbAl_3$ via MASHS reaction. Intermetallics. 2002;**10**:377-389

[125] Javel JF, Dirand M, Nassik FZ, Gachon JC. Real time X-ray diffraction study of the formation by SHS of the phases γ' and H in the ternary system Al Ni Ti. Le Journal de Physique IV. 1996;**6**(C2):229-234

[126] Javel JF, Dirand M, Kuntz JJ, Nassik FZ, Gachon JC. Real time X-ray diffraction study of the formation by SHS of the phases γ' and H in the ternary system Al-Ni-Ti. Journal of Alloys and Compounds. 1997;**247**(1–2):72-81

[127] Vaucher S, Stir M, Ishizaki K, Catala-Civera JM, Nicula R. Reactive synthesis of Ti-Al intermetallics during microwave heating in an E-field maximum. Thermochimica Acta. 2011;**522**(1):151-154

[128] Dubois S, Karnatak N, Beaufort MF, Bourdarias L, Renault PO, Vreb D. Influence of mechanical activation on TiC self propagating high temperature synthesis. Materials and Technologies. 2003;**18**(3):158-162

[129] Wong J, Larson EM, Waide PA, Frahm R. Combustion front dynamics in the combustion synthesis of refractory metal carbides and Di-borides using time-resolved X-ray diffraction. Journal of Synchrotron Radiation. 2006;**13**(4):326-335

[130] Larson EM, Wong J, Holt JB, Waide PA, Nutt G, Rupp B, Termninello LJ. Time resolved diffraction study of Ta-C solid combustion system. Journal of Materials Research. 1993;**8**(7): 1533-1541

[131] Jie-Cai H, Zhang XH, Wood JV. In-situ combustion synthesis and densification of TiC-xNi cermets. Materials Science and Engineering: A. 2000;**280**(2):328-333

[132] Riley DP, Kisi EH, Hansen TC, Hewat AW. Self-propagating high-temperature synthesis of Ti_3SiC_2: I, ultra-high-speed neutron diffraction study of the reaction mechanism. Journal of the American Ceramic Society. 2002;**85**(10):2417-2424

[133] Gauthier V, Cochepin B, Dubois S, Vrel D. Self-propagating high-temperature synthesis of Ti_3SiC_2: Study of the reaction mechanisms by time-resolved X-ray diffraction and infrared thermography. Journal of the American Ceramic Society. 2006;**89**(9):2899-2907

[134] Boutefnouchet H, Curfs C, Triki A, Boutefnouchet A, Vrel D. Self-propagating high-temperature synthesis mechanisms within the Ti-C-Ni system: A time resolved X-ray diffraction study. Powder Technology. 2012;**217**:443-450

[135] Mukasyan AS, Vadchenko SG, Khomenko IO. Combustion modes in the titanium-nitrogen system at low nitrogen pressures. Combustion and Flame. 1997;**111**(1–2):65-72

[136] Carole D, Fréty N, Paris S, Vrel D, Bernard F, Marin-Ayral RM. Investigation of the SHS mechanisms of titanium nitride by in situ time-resolved diffraction and infrared thermography. Journal of Alloys and Compounds. 2007;**436**(1):181-186

[137] Parkin IP, Pankhurst QA, Affleck L, Aguas MD, Kuznetsov MV. Self-propagating high temperature synthesis of $BaFe_{12}O_{19}$, $Mg_{0.5}Zn_{0.5}Fe_2O_4$ and $Li_{0.5}Fe_{2.5}O_4$; time resolved X-ray diffraction studies (TRXRD). Journal of Materials Chemistry. 2000;**11**(1):193-199

[138] Spiers H, Parkin IP, Pankhurst QA, Affleck L, Green M, Caruana DJ, Kuznetsov MV, Yao J, Vaughan G, Terry A, Kvick A. Self propagating high temperature synthesis of magnesium zinc ferrites ($Mg_xZn_{1-x}Fe_2O_3$): Thermal imaging and time resolved X-ray diffraction experiments. Journal of Materials Chemistry. 2004;**14**(7):1104-1111

[139] Kovalev DY, Shkiro VM, Ponomarev VI. Dynamics of phase formation during combustion of Zr and Hf in air. International Journal of Self-Propagating High-Temperature Synthesis. 2007;**16**(4):169-174

[140] Gras C, Gaffet E, Bernard F, Niepce JC. Enhancement of self-sustaining reaction by mechanical activation: Case of an Fe-Si system. Materials Science and Engineering: A. 1999;**264**(1):94-107

[141] Gras C, Bernsten N, Bernard F, Gaffet E. The mechanically activated combustion reaction in the Fe-Si system: In situ time-resolved synchrotron investigations. Intermetallics. 2002;**10**(3):271-282

[142] Gras C, Charlot F, Gaffet E, Bernard F, Niepce JC. In situ synchrotron characterization of mechanically activated self-propagating high-temperature synthesis applied in Mo-Si system. Acta Materialia. 1999;**47**(7):2113-2123

[143] Gras C, Gaffet E, Bernard F. Combustion wave structure during the MoSi2 synthesis by mechanically-activated self-propagating high-temperature synthesis (MASHS): In situ time-resolved investigations. Intermetallics. 2006;**14**(5):521-529

[144] Kachelmyer CR, Khomenko IO, Rogachev AS, Varma A. A Time-resolved X-ray diffraction study of Ti_5Si_3 product formation during combustion synthesis. Journal of Materials Research. 1997;**12**(12):3230-3240

[145] Riley DP, Oliver CP, Kisi EH. In-situ neutron diffraction of titanium silicide, Ti_5Si_3, during self-propagating high-temperature synthesis (SHS). Intermetallics. 2006;**14**(1):33-38

[146] Rupp B, Wong J, Holt JB, Waide P. The solid combustion synthesis of small $REBa_2Cu_3O_x$ samples (Re-Y, Er). The solid combustion synthesis of small $REBa_2Cu_3O_x$ samples (Re-Y, Er). Journal of Alloys and Compounds. 1994;**209**(1–2): 25-33

[147] Curfs C, Cano IG, Vaughan GBM, Turrillas X, Kvick Å, Rodrıguez MA. TiC-NiAl composites obtained by SHS: A time-resolved XRD study. Journal of the European Ceramic Society. 2002;**22**(7):1039-1044

[148] Terry AE, Vaughan GB, Kvick Å, Walton RI, Norquist AJ, O'Hare D. In situ time-resolved X-ray diffraction: The current state of the art. Synchrotron Radiation News. 2002;**15**(4):4-13

[149] Contreras L, Turrillas X, Vaughan GBM, Kvick Å, Rodrıguez MA. Time-resolved XRD study of TiC-TiB_2 composites obtained by SHS. Acta Materialia. 2004;**52**(16):4783-4790

[150] Mishra SK, Das SK, Ramachandrarao P, Belov DY, Mamyan SS. Self-propagating high-temperature synthesis of a zirconium diboride -alumina composite: A dynamic X-ray diffraction study. Philosophical Magazine Letters. 2004;**84**(1):41-46

[151] Carole D, Frety N, Paris S, Vrel D, Bernard F, Marin-Ayral RM. Microstructural study of titanium carbonitride produced by combustion synthesis. Ceramics International. 2007;**33**(8):1525-1534

[152] Mas-Guindal MJ, Turrillas X, Hansen T, Rodríguez MA. Time-resolved neutron diffraction study of Ti-TiC-Al_2O_3 composites obtained by SHS. Journal of the European Ceramic Society. 2008;**28**(15):2975-2982

[153] Kovalev DY, Prokudina VK, Ratnikov VI, Ponomarev VI. Thermal decomposition of TiH2: A TRXRD study. International Journal of Self-Propagating High-Temperature Synthesis. 2010;**19**(4):253-257

[154] Bazhin PM, Stolin AM, Alymov MI. Preparation of nanostructured composite ceramic materials and products under conditions of a combination of combustion and high-temperature deformation (SHS extrusion). Nanotechnologies in Russia. 2014;**9**(11–12): 583-600

[155] Manukyan KV, Shuck CE, Cherukara MJ, Rouvimov S, Kovalev DY, Strachan A, Mukasyan AS. Exothermic self-sustained waves with amorphous nickel. Journal of Physical Chemistry C. 2016;**120**(10):5827-5838

[156] LaGrange T, Campbell GH, Reed BW, Taheri M, Pesavento JB, Kim JS, Browning ND. Nanosecond time-resolved investigations using the in situ of dynamic transmission electron microscope (DTEM). Ultramicroscopy. 2008;**108**(11):1441-1449

[157] Shuck CE, Frazee M, Gillman A, Beason MT, Gunduz IE, Matouš K, Winarski R, Mukasyan AS. X-ray nanotomography and focused-ion-beam sectioning for quantitative three-dimensional analysis of nanocomposites. Journal of Synchrotron Radiation. 2016;**23**(4):990-996

[158] Lee SH, Lee JH, Lee YH, Shin DH, Kim YS. Effect of heating rate on the combustion synthesis of intermetallics. Materials Science and Engineering: A. 2000;**281**(1):275-285

[159] Yi HC, Petric A, Moore JJ. Effect of heating rate on the combustion synthesis of Ti-Al intermetallic compounds. Journal of Materials Science. 1992;**27**(24):6797-6806

[160] Moore JJ, Feng HJ. Combustion synthesis of advanced materials: Part I. Reaction parameters. Progress in Materials Science. 1995;**39**(4–5):243-273

[161] Yi HC, Moore JJ. Self-propagating high-temperature (combustion) synthesis (SHS) of powder-compacted materials. Journal of Materials Science. 1990;**25**(2):1159-1168

[162] Yum JT, Moon JT, Lee YH, Kim YS. Synthesis and microstructure control of TiAl via combustion synthesis. Metals and Materials International. 1995;**1**(1):19-27

Chapter 11

Ultrasound as a Metrological Tool for Monitoring Transesterification Kinetics

Raphaela M. Baêsso, Pâmella A. Oliveira,
Gabriel C. Moraes, André V. Alvarenga and
Rodrigo P.B. Costa-Félix

Additional information is available at the end of the chapter

http://dx.doi.org/10.5772/intechopen.70501

Abstract

Ultrasound has been widely used as a technological alternative way to analyse non-invasively an assortment of materials. It includes liquids with dissimilar physical characteristics, including mono- and multi-phasic mixtures, suspension formation and dissolution, in-line processing, among other practical applications. Regardless the huge spread of uses, so far ultrasound has not been proved to be able to quantify transesterification kinetics with a metrological approach. The aim of this chapter is to demonstrate that a properly designed ultrasonic experiment can be developed to identify remarkable stages of a transesterification reaction to produce biodiesel. The method was compared both with gas chromatography and hydrogen nuclear magnetic resonance (^{1}H NMR). For an in-line application, ultrasound has been proved to work properly as a monitoring tool for chemical reaction kinetics.

Keywords: ultrasound, metrology, chemical kinetics, monitoring, biodiesel production

1. Introduction

Currently, in the chemical, food, petrochemical and other industries, there is a considerable demand for measuring instruments that are able to characterize liquids with high sensitivity, robustness and precision. An instrument that is able to perform the process accurately, ranging from chemical reactions (production) to quality control (final product), is necessary. Due to the automation of processes, in-line measurements are increasingly being studied to ensure that the product is in conformity to technical requirements [1–5].

The competitiveness of companies for which the core business is the production of consumer goods is directly linked to their production process. Quality and productivity, once seen as dissociated elements, now are joined together, strongly impacting business competitiveness, improving process performance, product quality and reducing costs. In this way, the study of the kinetics of chemical processes allows the optimization of the process, avoiding the waste not only of raw materials, but also of energy and time [6].

Establishing the kinetics of a chemical reaction can be very complex depending on the chemical route. An in-depth study requires a series of experiments and simulations that can predict when the reaction has reached the optimal state. For reactions in which the converting mechanism is not well established, several kinetic models may be proposed, arising from countless reaction monitoring techniques. Thus, the combination of techniques, well described in the literature and easily accessible, allows a more precise conclusion of the object of study [6, 7].

To illustrate, let us take one of the chemical reactions responsible for biodiesel production, the transesterification reaction, disclosed in **Figure 1**.

Figure 1 discloses that an oil can react with a small chain alcohol in the presence of a catalyst and produce biodiesel. Biodiesel, also known as fatty acid methyl ester (FAME), is nothing more than a mixture of esters.

The transesterification reaction to produce biodiesel is reversible and, therefore, usually works with excess of alcohol. A typical proportion is 6:1 (alcohol:oil) ratio, so that the equilibrium displacement is forced towards the conversion of biodiesel. However, the oil and alcohol are not miscible, establishing two phases. In this stage, when there are two phases within the mixture, one can say that the mass transfer controls the kinetics of the reaction. Nevertheless, as biodiesel (methyl ester) is formed, it works as a co-solvent, facilitating the miscibility. Upon reaching the homogeneity of the system, the chemical reaction starts to control the system. A reaction in which an exchange of mechanism occurs makes difficult to define the kinetics and, consequently, the establishment of optimal reaction conditions [7–12].

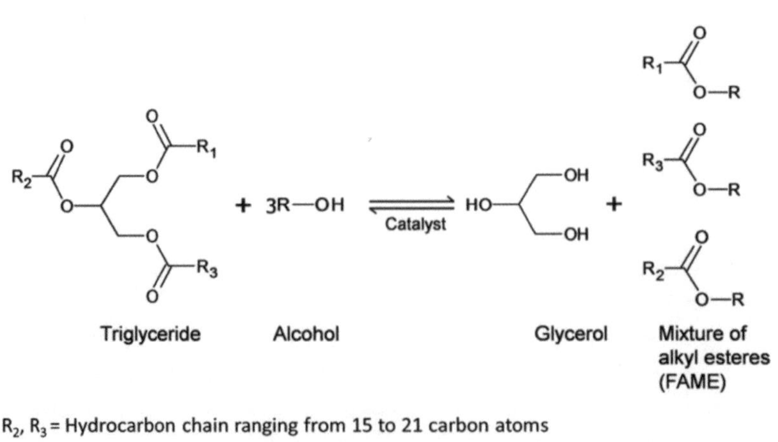

Figure 1. Transesterification of vegetable oil for biodiesel production.

Within this chapter, we will disclose the recent outcomes of a research in which ultrasound has been used as a way of monitoring the progress of the reaction of transesterification of soybean oil with methanol in the presence of KOH as basic catalyst. The pulse/echo technique was used to monitor acoustic velocity throughout the reaction, composed as an in-line scheme. The results were related to the reference method based on gas chromatography (EN 14103 standard) and to the ^{1}H NMR technique [13–18].

2. Fundamental of the transesterification reaction

Let us consider the reaction TAG + 6 MeOH $\rightarrow$ 3 FAME + G, where TAG (triglyceride) is soybean oil, MeOH is methanol, FAME is the ester mixture and G is glycerol. To simplify the study, there are some considerations before starting the process of monitoring the transesterification reaction by ultrasound. As there is excess alcohol, we can consider TAG as a limiting reagent, i.e., when it is totally consumed the reaction will finish. With the excess of methanol, the reaction becomes irreversible, which means that the whole equilibrium of the reaction is displaced to produce biodiesel. Another important consideration is to admit a batch reactor in which reagents are mixed at the beginning of the reaction, without any inlet or outlet flow of reagents further than those mixed at time 0.

Thus, we determine the amount of TAG consumed (N_{TAG_c}), quantified in mol, through Eq. (1), in which N_{TAG_0} represents the amount of TAG inserted in the reactor at time 0 (in mol), and X is the relationship between the TAG that reacted with the TAG inserted into the reactor at time 0.

$$N_{TAG_c} = N_{TAG_0} \cdot X \tag{1}$$

The TAG in a given time (N_{TAG_t}) is obtained by the difference of the TAG inserted in the reactor and the TAG consumed, according to Eq. (2) (all values expressed in mol).

$$N_{TAG_t} = N_{TAG_0} - N_{TAG_0} \cdot X \tag{2}$$

Thus, Eqs. (1) and (2) lead to Eq. (3), as depicted elsewhere [6, 9].

$$N_{TAG_t} = N_{TAG_0}(1 - X) \tag{3}$$

From stoichiometry, one can derive to $C_{TAG} = \frac{N_{TAG}}{V}$, and for liquids $V = V_0$ (volume is constant), evolving to Eq. (4).

$$C_{TAG} = C_{TAG_0}(1 - X) \tag{4}$$

The disappearance of the TAG must be accompanied by the appearance of the FAME. However, a question arises: how the actual concentration of FAME throughout the reaction can be assessed? Complementary, one could ask how to know that in fact the reaction has ended and it has already reached the maximum conversion? There are several techniques proposed in the literature, but few can make this determination in an in-line scheme with an accurate way without demanding exorbitant expenses. To sort out that drawback, ultrasound methods emerge as a tool capable of assisting in the biodiesel manufacturing process.

3. Ultrasound as a tool for liquid characterization

Ultrasound is a mechanical wave that propagates in fluids or solid materials at frequencies greater than 20 kHz, i.e., out of the audible range for healthy humans [19–25].

An often used ultrasonic measurement method consists on a pulse/echo arrangement. Basically, it consists on

$$\text{EMISSION} \rightarrow \text{PROPAGATION} \rightarrow \text{REFLECTION} \rightarrow \text{PROPAGATION} \rightarrow \text{RECEPTION}$$

Two important quantities are easily assessed from the pulse/echo ultrasonic measurement method: time of flight and pulse (or signal) amplitude. Both are measured after the reception of the ultrasound wave. Whenever an acoustic impedance mismatch occurs, the ultrasonic wave is partially reflected in the discontinuity boundary. The amount of reflection depends on the acoustic impedance difference between the two media, due to what is called the reflection coefficient of the interface. In a typical pulse/echo experimental set-up in sonochemistry, the propagation medium is fluid and the reflection takes place in an interface with a solid object, generically denominated reflecting target. Similarly, the liquid-air interface is a reflecting target, as well.

Throughout the propagation, other physical phenomena diminish the ultrasonic amplitude due to different mechanisms. Mainly, scattering and absorption are in charge for ultrasonic attenuation, mitigating the capability of free propagation. All those phenomena are natural and unavoidable. Nevertheless, a proper experimental ultrasonic set-up will either concern on its quantification, or will deal with other quantities that are not undesirably affected.

The speed of sound is a quantity that is not related to attenuation phenomena, or at least is not the case in a linear range of frequencies and in infinite-like three-dimensional propagation medium, even if there is a constraint in one dimension. In the linear range of ultrasonic propagation, the sound velocity in any determined medium or material varies as a function of the temperature, density and viscosity. As a matter of fact, those quantities are not absolutely correlated to each other, what makes the establishment of a mathematical function a virtually unrealisable task for complex mixtures of fluids. For monophasic simple liquids, such as pure water or hydro carbonates, it is easier to define a function relating those quantities, but it is not the case for a transesterification process.

To assess the speed of sound, the typical approach is to measure the time of flight of an ultrasonic pulse within a vessel with a pre-determined distance from the surface to the emitting ultrasonic transducer and a properly designed reflecting target. It is the so-called pulse/echo experimental method. Materials in different macrophysics states transmit ultrasonic waves with different velocities. In general, but not in a universal way, the more rigid is a

http://dx.doi.org/10.5772/intechopen.70501

material, the faster the ultrasonic wave will propagate within it. It is important to keep in mind that the ultrasonic velocity changes significantly with temperature [26–29].

Mathematically, the speed of sound is computed dividing the distance travelled by the pulse by the time spent to travel it (time of flight), as disclosed in Eq. (5).

$$\upsilon = \frac{2 \cdot \Delta s}{t} \tag{5}$$

Here, Δs is the distance separating the ultrasonic surface and the reflecting interface (the travelling distance is twice this value) and t is the time required for the ultrasonic pulse to transpose that distance and return to the transducer. This process can be repeated many times, depending on the attenuation and the distance from the transducer and the reflecting surface. After each subsequent reflection, the pulse amplitude will decrease, as a consequence of attenuation. The multiple reflections will remain until the sound energy is completely absorbed in the process. **Figure 2** exhibits that multi-reflection behaviour.

While planning the experimental set-up for the pulse/echo method, one must be aware about the absorption of the liquid under investigation, as well as the distance between the transducer and the reflecting surface. The pulse frequency plays a key role, as ultrasonic attenuation is exponentially proportional to the frequency. In general, water is used as reference once its behaviour both for attenuation and ultrasonic velocity are very well known [26–29].

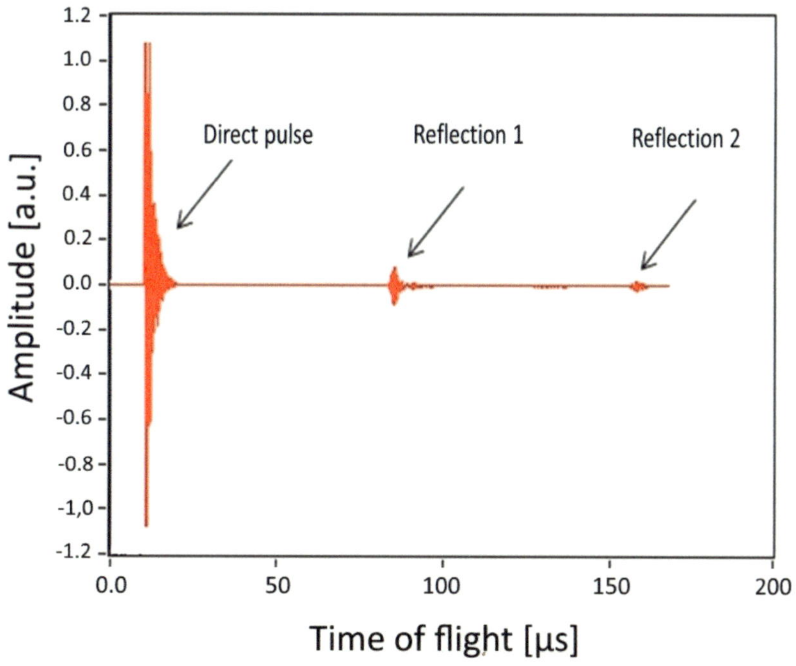

Figure 2. Ultrasonic pulse and reflections.

4. Validation of the experimental ultrasonic method

In previous studies, the value of the propagation speed in soybean oil in a range of 20–50°C was determined [20]. For the practical application presented within this paper, all reactions were performed at 40°C. At this temperature, soybean oil has a velocity of 1418.3 m s^{-1}, with expanded uncertainty (p = 0.95) U_{exp} = 5.2 m s^{-1} [21]. As soon as the oil starts to react with methanol and the catalyst, variation on the speed of sound will indicate that something is happening within the medium. Despite it is easy to measure speed of sound, it is not trivial to relate this variation with anything that is going on in the reaction. The chemical kinetics is not directly assessed, unless some methodological study is conducted. That was the case, insofar we conducted an experimental method validation. The idea was to compare the speed of sound measured throughout the transesterification process with a quantification of the reaction stoichiometric situation at different moments. The worldwide accepted reference method for determination of ester content is based on gas chromatography (GC), according to the standard EN 14103. However, this method, besides being time-consuming, it is not applicable in the process line and demands expensive equipment, supplies, and specific technical training. Thus, less costly methods have emerged as an alternative for determining the conversion, as is the case with ^{1}H NMR. Despite it is not a cheap technique, it is much less time-consuming than the GC analysis [13, 14, 21, 22, 30–32].

In establishing parameters for the reactions that will be analysed, an isothermal batch reactor (T = 40°C) is chosen. The validation experiment was restricted to two concentrations for the catalyst (0.2% and 1.5% w/w) and two mechanical stirring rotational speeds (200 and 520 rpm). The reaction time was set to a limit of 40 min.

There are several studies that propose equations that take into account the number of hydrogens present in the molecules consumed (TAG) in relation to the number of hydrogens present in the formed molecule (FAME). **Figures 3** and **4** disclose the ^{1}H NMR spectra for the pure soybean oil and the biodiesel made from this oil, respectively.

The formation of methyl ester (methylic biodiesel) can be noticed by the appearance of the signal of the methylic hydrogen from the methoxyl group at 3.7 ppm (chemical shift represented per B in **Figure 4**), while occurs the disappearance of the methylene hydrogens from glycerol in the triacylglycerol from 4.0 to 4.4 ppm (chemical shift represented per B in **Figure 3**). Eq. (6) presents a method described in the literature [14] used to determine the conversion of TAG.

$$X_{TAG}(\%) = 100 \cdot \left(\frac{2A_1}{3A_2}\right) \tag{6}$$

Here, X_{TAG} (%) is the amount of TAG that has been converted into biodiesel, A_1 and A_2 are areas of the methylic hydrogens (δ = 3.7 ppm) of methoxyl group, the methyl ester and the glycerol methylene hydrogens (δ = 4–4.4 ppm), respectively. Calculation of the ^{1}H NMR conversion was compared to the reference method (GC) for the two reactions with 200 rpm (see **Figure 5**).

http://dx.doi.org/10.5772/intechopen.70501

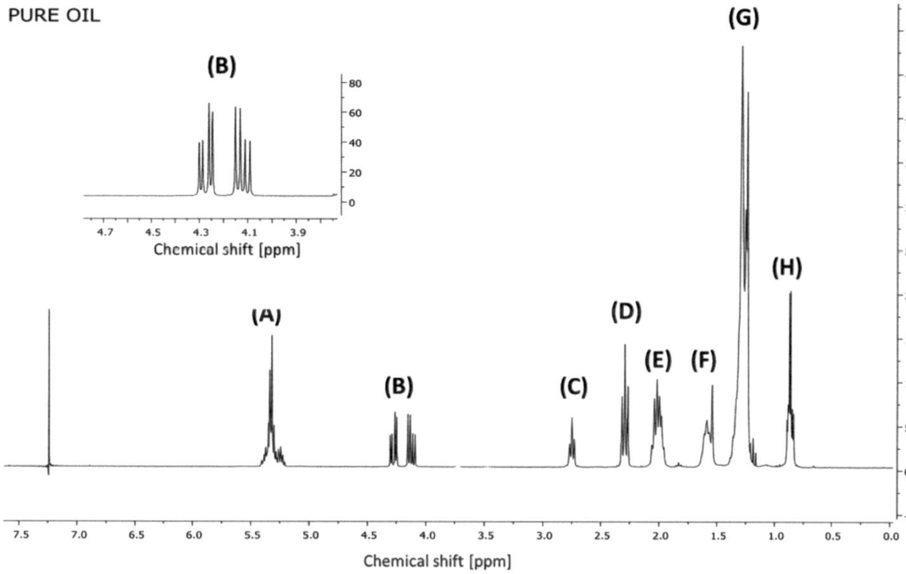

Figure 3. ^{1}H NMR spectra for pure soybean oil: (A) HC=CH; (B) CH_2—O; (C) C=C—CH_2—C=C; (D) CH_2—C=O; (E) CH_2—C=C; (F) CH_2—C—C=O; and (G) —CH_2—; (H) CH_3.

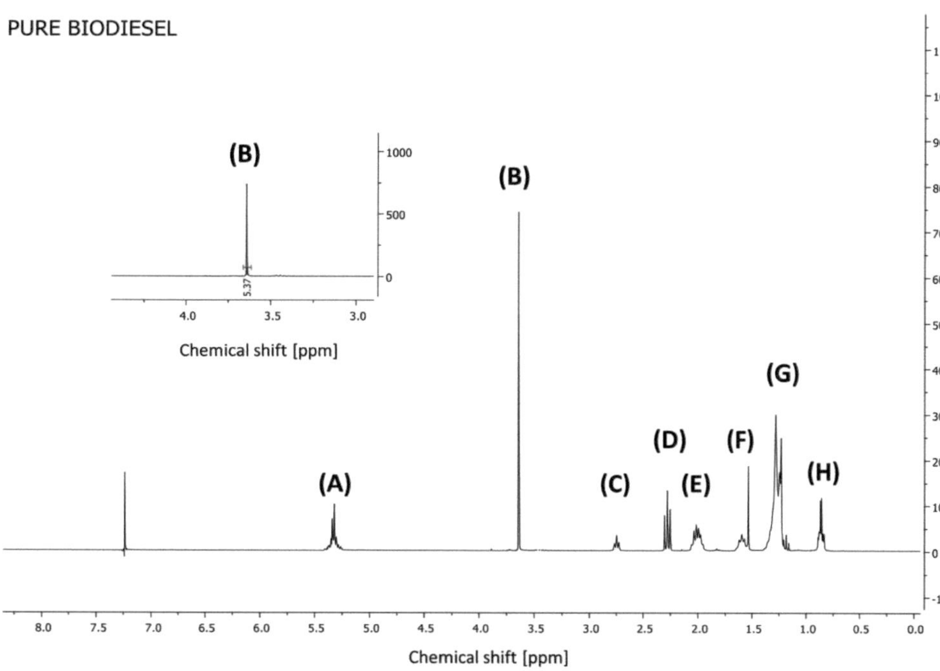

Figure 4. ^{1}H NMR spectra of methyl biodiesel obtained on the homogeneous transesterification of soybean oil: (A) HC=CH; (B) CH_3—O; (C) C=C—CH_2—C=C; (D) CH_2—C=O; (E) CH_2—C=C; (F) CH_2—C—C=O; and (G) —CH_2—; (H) CH_3.

Figure 5 shows that the ^{1}H NMR method is very similar to the reference method (GC). After this straightforward validation, comparing both technics, we used ^{1}H NMR to assess the conversion of analytical curves as described in Eq. (6) for the four reactions (see **Figure 6**).

As disclosed in **Figure 6**, one can note that each reaction reaches the maximum conversion at a given moment. But how to know during the reaction that the maximum conversion has already been reached and there is no longer any need to continue the process? NMR analyses, as well as GC, require the sample to be pure, which means free of other substances that may interfere with the analysis. In this way, ultrasonic monitoring stands out, being able to determine the maximum point of the reaction even in the presence of excess reagents and by-products. **Figures 7** and **8** depict a set of results for all chemical routes employed in the present study.

Figure 7 shows that each reaction has a propagation velocity configuration. They all start with a value close to the pure soybean oil velocity. However, during the reaction time, the speed of

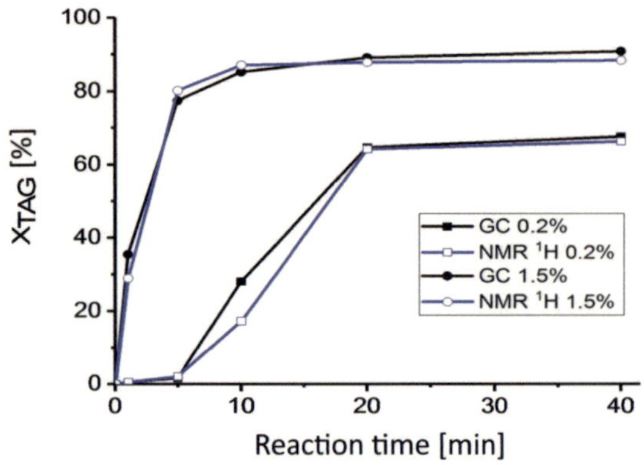

Figure 5. Variation of biodiesel conversion during the homogeneous transesterification of soybean oil with methanol and 200 rpm of mechanical stirring.

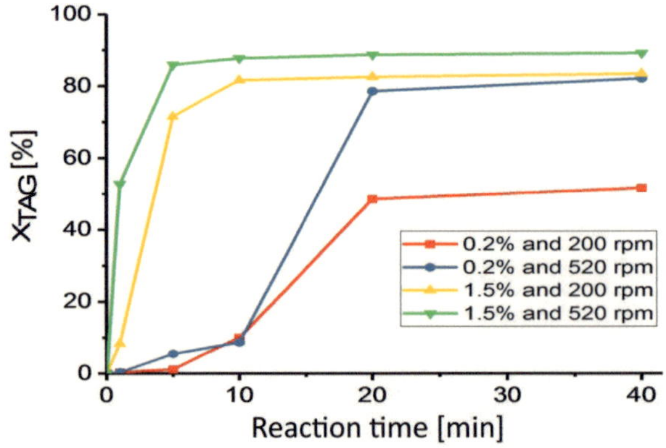

Figure 6. Conversion rate calculated by ^{1}H NMR for four transesterification reactions of soybean oil using KOH as the basic catalyst.

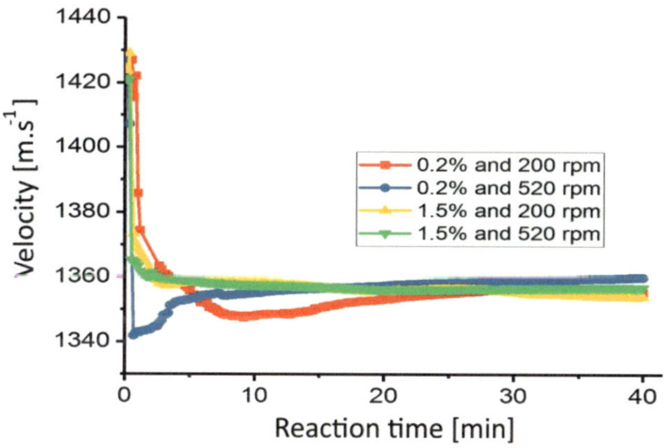

Figure 7. Variation of the propagation velocity along the transesterification reaction of soybean oil with methanol in the presence of KOH.

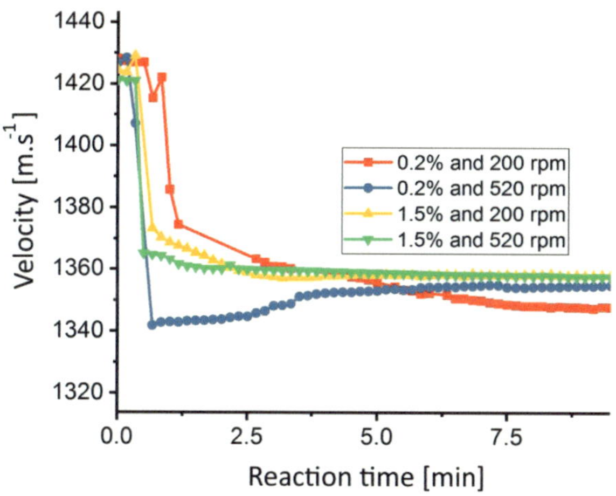

Figure 8. First 7.5 min of the variation of the propagation velocity along the transesterification of soybean oil with methanol in the presence of KOH.

sound decreases until stabilized, demonstrating that the maximum conversion was reached. This variation can be better observed in **Figure 8**, which is a zoomed part of **Figure 7** restricted to the first few minutes. Considering that the idea is to obtain pure biodiesel, independent of the reaction conditions, it is quite natural that the final velocities (when the highest concentration of biodiesel is present) are close to each other.

5. Looking into the results in details

Let us analyse the effect of each parameter (catalyst and rotation) on the final biodiesel conversion.

5.1. Analysis of the effect of catalyst concentration

The presence of catalyst helps to accelerate the reaction, reducing the activation energy required to start it. Thus, catalyst concentration is one of the main factors that can affect the reaction kinetics.

In order to analyse the effect of the variation of the catalyst concentration, it is necessary to separately analyse different rotation values.

For an initial concentration of soybean oil equal to 0.83 mol L^{-1} and by the Eqs. (4) and (6), the consumption of TAG and conversion into FAME were calculated. The velocity (υ) was calculated according to Eq. (5).

Table 1 and **Figure 9** disclose the results for the reactions with 200 rpm of rotational speed.

	0.2% and 200 rpm			1.5% and 200 rpm		
Time [min]	X_{TAG} [%]	C_{TAG} [mol L^{-1}]	υ [m s^{-1}]	X_{TAG} [%]	C_{TAG} [mol L^{-1}]	υ [m s^{-1}]
0	0	0.83	1428.1	0	0.83	1424.3
1	0.3	0.83	1385.7	8.3	0.75	1368.7
5	1.1	0.82	1355.8	71.5	0.24	1358.0
10	10.0	0.75	1347.9	81.6	0.15	1356.1
20	48.5	0.43	1353.2	82.5	0.15	1355.4
40	51.6	0.40	1355.2	83.4	0.14	1353.9

Table 1. Results for reactions at 200 rpm.

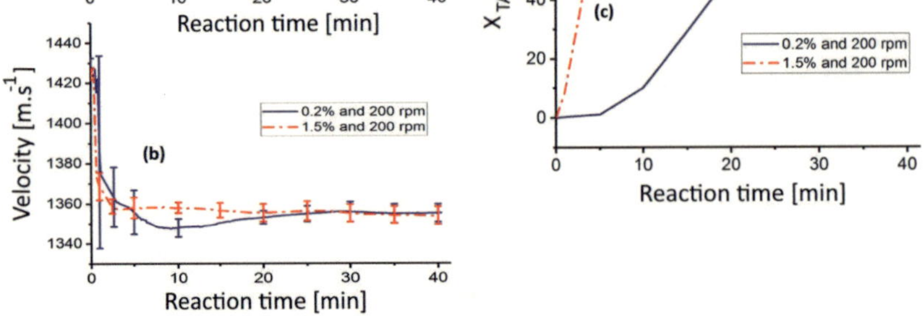

Figure 9. Variation of (a) TAG concentration, (b) propagation velocity and (c) TAG conversion for the reactions with 200 rpm of rotation.

Looking at **Table 1** and **Figure 9**, one can note that there is a large difference in C_{TAG} and υ decay between both reactions, what reflects in the conversion rate (X_{TAG}). The uncertainty bars calculated for the velocity are disclosed to determine the accuracy of the study [33]. The behaviour observed in the reaction with 0.2% KOH and 200 rpm presents a slow decay in the concentration of TAG in the first 10 min of reaction, which is expected for reactions in which mechanism exchange occurs. That region on the graphics of **Figure 9** (first 10 minutes of reaction) is in which the rate of consumption and conversion rate are slow, and it occurs due to the low miscibility between alcohol and oil. Thus, as the methyl ester is produced, the chemical reaction starts to control the kinetics of the reaction and, therefore, a large jump between 10 and 20 min of reaction is observed. After 20 min of reaction, it is clearly noticeable that there is stability between the conversion values and propagation velocity. This result indicates that the ultrasound can determine the maximum point of conversion even in the presence of secondary substances (by-products).

On the other hand, when the reaction with 1.5% KOH and 200 rpm is on focus, yet in **Figure 9**, it is observed an expressive consumption of TAG in the first 10 min of reaction. After that time, the stability in the values of the conversion as well as in the propagation velocity is evident. Reactions like that, in which there is a rapid conversion, the region controlled by mass transfer can be considered insignificant. Here, it is observed that for reactions with 200 rpm stirring, the increase in catalyst concentration not only accelerates the transesterification process but also increases the final conversion as well, and consequently decreases the remaining TAG in the reaction medium.

Let us check if the same will occur analysing the results presented in **Table 2** and **Figure 10**, in which the rotation speed was increased to 520 rpm.

For the reaction with 0.2% of KOH and 520 rpm of stirring, it is noticeable that there is a slow decrease in the TAG concentration in the first 10 min and stability of the values after 20 min. On the other hand, the reaction with 1.5% of KOH and 520 rpm reaches the maximum conversion as fast as 5 min after the reaction had begun. In the same way as observed in **Figure 9**, the mass transfer controls the start of the reaction for the reaction with 0.2% of catalyst, independently of the rotational speed employed. With the increase of KOH concentration by 7.5 times, there is a 300% decrease in reaction time for the reaction with 520 rpm of

	0.2% and 520 rpm			1.5% and 520 rpm		
Time [min]	X_{TAG} [%]	C_{TAG} [mol L^{-1}]	υ [m s^{-1}]	X_{TAG} [%]	C_{TAG} [mol L^{-1}]	υ [m s^{-1}]
0	0	0.83	1426.8	0	0.83	1421.3
1	0.3	0.81	1342.9	52.8	0.65	1363.2
5	5.5	0.78	1353.0	85.9	0.12	1358.8
10	8.7	0.76	1354.0	87.7	0.10	1357.6
20	78.6	0.22	1357.3	88.7	0.09	1356.3
40	82.1	0.18	1360.0	89.1	0.09	1356.9

Table 2. Results for reactions at 520 rpm.

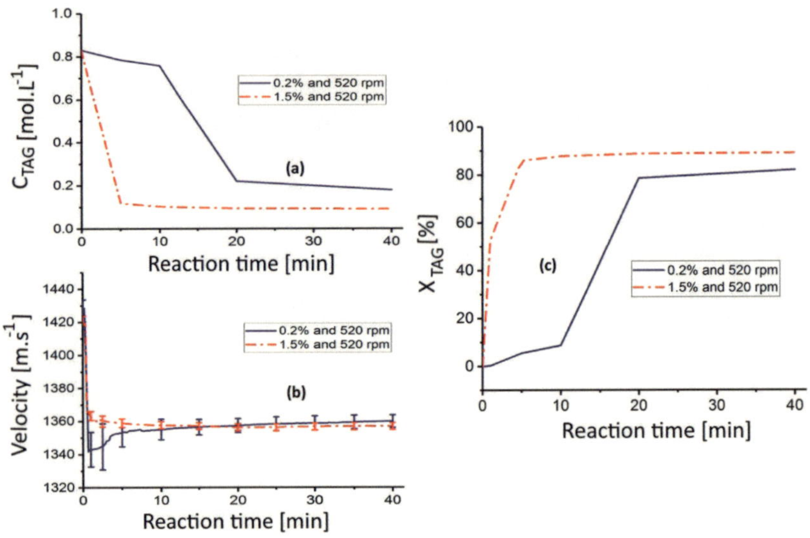

Figure 10. Variation of (a) TAG concentration, (b) propagation velocity and (c) TAG conversion for the reactions with 520 rpm of rotation.

rotation. However, the maximum values reached by the conversion for the two reactions described in **Figure 10** are very close (89 and 82%).

5.2. Analysis of the effect of system stirring

If the same concentration of catalyst is used, would the stirring of the system interfere with the kinetics of the reaction? The answer is: Surely enough!

Firstly, the lowest catalyst concentration, 0.2% KOH, will be analysed. The impact due the change in rotational speed will be variable in those reactions. **Table 3** and **Figure 11** show the results for the reactions with 0.2% catalyst.

From **Table 3** and **Figure 11**, we note that the two reactions with 0.2% of catalyst need the same reaction time to reach their maximum conversion, regardless of the rotation applied to the

	0.2% and 200 rpm			0.2% and 520 rpm		
Time [min]	X_{TAG} [%]	C_{TAG} [mol L^{-1}]	υ [m s^{-1}]	X_{TAG} [%]	C_{TAG} [mol L^{-1}]	υ [m s^{-1}]
0	0	0.83	1428.1	0	0.83	1426.8
1	0.3	0.83	1385.7	0.3	0.81	1342.9
5	1.1	0.82	1355.8	5.5	0.78	1353.0
10	10.0	0.75	1347.9	8.7	0.76	1354.0
20	48.5	0.43	1353.2	78.6	0.22	1357.3
40	51.6	0.40	1355.2	82.1	0.18	1360.0

Table 3. Results for 0.2% (w/w) of KOH reactions.

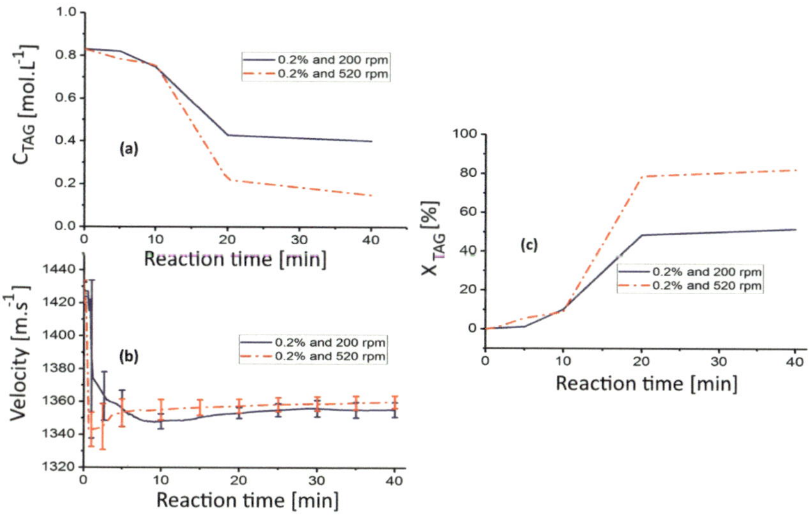

Figure 11. Variation of (a) TAG concentration, (b) propagation velocity and (c) TAG conversion for the reactions with 0.2% (w/w) of KOH.

system. However, there is a significant increase in the conversion value from 52% (200 rpm) to 82% (520 rpm), which shows that for reactions like those, with low catalyst concentration, the rotation does not interfere in the reaction velocity, but at the maximum conversion value.

And what happens increasing 7.5 times the concentration of the catalyst? Does this pattern hold? **Table 4** and **Figure 12** show the answers.

Increasing the concentration of KOH clearly increases the rate of TAG consumption and, consequently, formation of FAME. While the reaction with 200 rpm reaches the maximum conversion and the equilibrium with 10 min of reaction, the reaction at 520 rpm only requires half the time, 5 min. However, despite the decrease in time, we observed that the values for the maximum conversion are very close.

	1.5% and 200 rpm			1.5% and 520 rpm		
Time [min]	X_{TAG} [%]	C_{TAG} [mol L^{-1}]	υ [m s^{-1}]	X_{TAG} [%]	C_{TAG} [mol L^{-1}]	υ [m s^{-1}]
0	0	0.83	1424.3	0	0.83	1421.3
1	8.3	0.75	1368.7	52.8	0.65	1363.2
5	71.5	0.24	1358.0	85.9	0.12	1358.8
10	81.6	0.15	1356.1	87.7	0.10	1357.6
20	82.5	0.15	1355.4	88.7	0.09	1356.3
40	83.4	0.14	1353.9	89.1	0.09	1356.9

Table 4. Results for 1.5% (w/w) of KOH reactions.

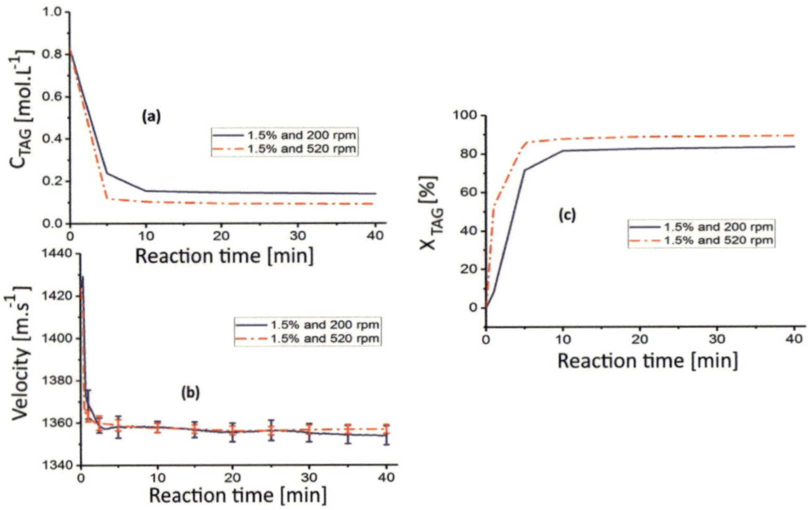

Figure 12. Variation of (a) TAG concentration, (b) propagation velocity and (c) TAG conversion for the reactions with 1.5% (w/w) of KOH.

Thus, an economic analysis is necessary to evaluate to what extent an increase of the rotation used, or of the catalyst concentration, to the detriment of the reduction in the reaction time, is economically feasible. But the use of ultrasound as a tool for monitoring chemical reactions has been shown to be efficient [33].

6. Final remarks

Ultrasound is widely used for more than a century for diagnosis applications. The state of the technology is vast on applications for non-destructive testing and biomedical equipment. The use of ultrasound in chemistry is more common as a tool to accelerate reactions or enhance the performance of established methods. However, the technology is not so widely developed and spread around regarding the use of ultrasound as a monitoring tool for chemical reactions.

As a tool, ultrasound is remarkably simple to use. Nevertheless, one must be aware that the apparent straightforwardness undercovers a complex physical process that takes place in the generation, propagations, reflection and reception of ultrasound in both the transmit/receive and pulse/echo approaches. Unless an experiment is carefully designed, carried out, and analysed, the outcome of any ultrasonic proposed method could be of no technical usefulness.

In the present chapter, the use of an ultrasound pulse/echo scheme was validated as a monitoring procedure of the transesterification kinetics of soybean oil into biodiesel. The sensibility of the method was good enough to compare different catalyst concentrations (0.2 and 1.5%) and different rotational speed of mechanical stirring (200 and 520 rpm). The comparison was done using as gold standard the gas chromatography and ^{1}H RMN. The validation leads to quite interesting outcomes.

It was possible to observe purely from the ultrasonic velocity measurement that the faster the mechanical stirring acts, the faster is the transesterification kinetics. Moreover, it is possible to identify the elapsed time when the reaction reaches its maximum possible conversion, dictated by the amount of catalyst. For all cases, ultrasonic monitoring has disclosed a causal relation to the gold standard analytical methods.

Author details

Raphaela M. Baêsso, Pâmella A. Oliveira, Gabriel C. Moraes, André V. Alvarenga and Rodrigo P.B. Costa-Félix*

*Address all correspondence to: rpfelix@inmetro.gov.br

Laboratory of Ultrasound – National Institute of Metrology, Quality and Technology (Inmetro), Duque de Caxias, RJ, Brazil

References

[1] Greenwood MS, Bamberger JA. Measurement of viscosity and shear wave velocity or slurry for on-line process control. Ultrasonics. 2002;**39**(9):623-630. DOI: https://doi.org/10.1016/S0041-624X(02)00372-4

[2] Saggin R, Coupland JN. Oil viscosity measurement by ultrasonic reflectance. Journal of the American Oil Chemists' Society. 2001;**78**(5):509-511. DOI: https://doi.org/10.1007/s11746-001-0294-z

[3] Saggin R, Coupland JN. Concentration measurement by acoustic reflectance. Journal of Food Science. 2001;**66**(5):681-685. DOI: https://doi.org/10.1111/j.1365-621.2001.tb04621.x

[4] Richard R, Li Y, Dubreuil B, Thiebaud-Roux S, Prat L. On-line monitoring of the transesterification reaction between triglycerides and ethanol using near infrared spectroscopy combined with gas chromatography. Bioresource Technology. 2011;**102**(2):6702-6709. DOI: https://doi.org/10.1016/j.biortech.2011.03.111

[5] Dubé MA, Zheng S, McLean DD, Kates M. A comparison of attenuated total reflectance-FTIR spectroscopy and GPC for monitoring biodiesel production. Journal of the American Oil Chemists' Society. 2004;**81**(6):599-603. DOI: https://doi.org/10.1007/s11746-006-0948-x

[6] Fogler HS. Elementos de Engenharia das Reações Químicas. 4th ed. São Paulo, SP: Editora LTC; 2009

[7] Kusdiana D, Saka S. Kinetics of transesterification in rapeseed oil to biodiesel fuel as treated in supercritical methanol. Fuel. 2001;**80**:693-698. DOI: https://doi.org/10.1016/S0016-2361(00)00140-X

[8] Galvan D, Orives JR, Coppo RL, Silva ET, Angilelli KG, Borsato D. Determination of the kinetics and thermodynamics parameters of biodiesel oxidation reaction obtained from an optimized mixture of vegetable oil and animal fat. Energy Fuels. 2013;**27**(11):6866-6871. DOI: https://doi.org/10.1021/ef401927x

[9] Vicente G, Martínez M, Aracil J, Esteban A. Kinetics of sunflower oil methanolysis. Industrial & Engineering Chemistry Research. 2005;**44**(15):5447-5454. DOI: https://doi.org/10.1021/ie040208j

[10] Stamenkovic OS, Todorovic ZB, Lazic LM, Veljkovic VB, Skala DU. Kinetics of sunflower oil methanolysis at low temperatures. Bioresource Technology. 2008;**99**(5):1131-1140. DOI: https://doi.org/10.1016/j.biortech.2007.02.028

[11] Clark WM, Medeiros NJ, Boyd DJ, Snell JR. Biodiesel transesterification kinetics monitored by pH measurement. Bioresource Technology. 2013;**136**:771-774. DOI: https://doi.org/10.1016/j.biortech.2013.03.089

[12] Tapanes NCO, Aranda DAG, Carneiro JWM, Antunes OAC. Transesterification of Jatropha curcas oil glycerides: Theoretical and experimental studies of biodiesel reaction. Fuel. 2008;**87**(10-11):2286-2295. DOI: https://doi.org/10.1016/j.fuel.2007.12.006

[13] EUROPEAN COMMITTEE FOR STANDARDIZATION. CSN EN 14103 – Fat and oil derivatives – Fatty Acid Methyl Esters (FAME) – Determination of ester and linolenic acid methyl ester contents. Brussels: CEN. 2011

[14] Gelbard G, Brès O, Vargas RM, Vielfaure F, Schuchardt UF. 1H nuclear magnetic resonance determination of the yield of the transesterification of rapeseed oil with methanol. Journal of the American Oil Chemists' Society. 1995;**72**:1239-1241. DOI: https://doi.org/10.1007/BF02540998

[15] Nehdi IA, Sbihi HM, Al-Resayes S.I. Rhazya stricta Decne seed oil as an alternative, nonconventionalfeedstock for biodiesel production. Energy Conversion and Management. 2014;**81**:400-406. DOI: http://dx.doi.org/10.1016/j.enconman.2014.02.038

[16] Kumar R, Bansal V, Patel VMB, Sarpal AS. 1H Nuclear magnetic resonance (NMR) determination of the iodine value in biodiesel produced from algal and vegetable oils. Energy Fuels. 2012;**26**:7005-7008. DOI: https://doi.org/10.1021/ef300991n

[17] Kouame S-D B, Perez J, Eser S, Benesi A. 1H-NMR monitoring of the transesterification process of Jatropha oil. Fuel Processing Technology. 2012;**97**:60-64. DOI: https://doi.org/10.1016/j.fuproc.2012.01.006

[18] Farobie O, Matsumura Y. A comparative study of biodiesel production using methanol, ethanol, and tert-butyl methyl ether (MTBE) under supercritical conditions. Bioresource Technology. 2015;**191**:306-311. DOI: https://doi.org/10.1016/j.biortech.2015.04.102

[19] Killner MHM, Rohwedder JJR, Pasquini C. A PLS regression model using NIR spectroscopy for on-line monitoring of the biodiesel production reaction. Fuel. 2011;**90**:3268-3273. DOI: https://doi.org/10.1016/j.fuel.2011.06.025

[20] Oliveira PA, Baêsso RM, Morais GC, Alvarenga AV, Costa-Félix RPB. Speed of sound as a function of temperature for ultrasonic propagation in soybean oil. Journal of Physics Conference Series (Online). 2016;**733**:012040. DOI: http://dx.doi.org/10.1088/1742-6596/733/1/012040

[21] Baêsso RM, Oliveira PA, Morais GC, Alvarenga AV, Costa-Félix RPB. Use of ultrasound to monitor physical properties of soybean oil. Journal of Physics Conference Series (Online). 2016;**733**:012042. DOI: https://doi.org/10.1088/1742-6596/733/1/012042

[22] Reddy SR, Titu D, Chadha A. A Novel method for monitoring the transesterification reaction of oil in biodiesel production by estimation of glycerol. Journal of the American Oil Chemists' Society. 2010;**87**(7):747-754. DOI: https://doi.org/10.1007/s11746-010-1549-2

[23] Koc AB. Ultrasonic monitoring of glycerol settling during transesterification of soybean oil. Bioresource Technology. 2009;**100**:19-24. DOI: https://doi.org/10.1016/j.biortech.2008.05.037

[24] Georgogianni KG, Kontominas MG, Pomonis PJ, Avlonitis D, Gergis V. Conventional and in situ transesterification of sunflower seed oil for the production of biodiesel. Fuel Processing Technology. 2008;**89**(5):503-509. DOI: https://doi.org/10.1016/j.fuproc.2007.10.004

[25] Figueiredo MK-K, Alvarenga AV, Costa-Félix RPB. Ultrasonic attenuation and sound velocity assessment for mixtures of gasoline and organic compounds. Fuel. 2017;**191**:170-175. DOI: https://doi.org/10.1016/j.fuel.2016.11.076

[26] Alouache B, Khechena FK, Lecheb F, Boutkedjirt T. Characterization of olive oil by ultrasonic and physico-chemical methods. Physics Procedia. 2015;**70**:106-1065. DOI: https://doi.org/10.1016/j.phpro.2015.08.226

[27] Meier K, Hawary AE. Measurements of the speed of sound in liquid and supercritical ethane. Fluid Phase Equilibria. 2016;**418**:125-132. DOI: https://doi.org/10.1016/j.fluid.2015.11.019

[28] Del Grosso VA, Mader CW. Speed of sound in pure water. The Journal of the Acoustical Society of America. 1972;**52**:1442. DOI: http://dx.doi.org/10.1121/1.1913258

[29] Monteiro MR, Ambrozin ARP, Lião LM, Ferreira AG. Critical review on analytical methods for biodiesel characterization. Talanta. 2008;**77**(2):593-605. DOI: https://doi.org/10.1016/j.talanta.2008.07.001

[30] Nagy M, Foston M, Ragauskas AJ. Rapid quantitative analytical tool for characterizing the preparation of biodiesel. The Journal of Physical Chemistry, A. 2010;**114**(11):3883-3887. DOI: https://doi.org/10.1021/jp906543g

[31] Sáez-Bastante J, Ortega-Román C, Pinzi S, Lara-Raya FR, Leiva-Candia DE, Dorado MP. Ultrasound-assisted biodiesel production from Camelina sativa oil. Bioresource Technology. 2015;**185**:116-124. DOI: http://dx.doi.org/10.1016/j.biortech.2015.02.090

[32] Joint Committee for Guides in Metrology. Evaluation of Measurement data – Guide to the Expression of Uncertainty in Measurement – JCGM 100:2008 [Internet]. 2008. Available from: http://www.bipm.org/en/publications/guides/gum.html [Accessed: 13 January 2017]

[33] Gaeta-Bernardi A, Parente V. Organic municipal solid waste (MSW) as feedstock for biodiesel production: A financial feasibility analysis. Renewable Energy. 2016;**86**:1422-1432. DOI: https://doi.org/10.1016/j.renene.2015.08.025